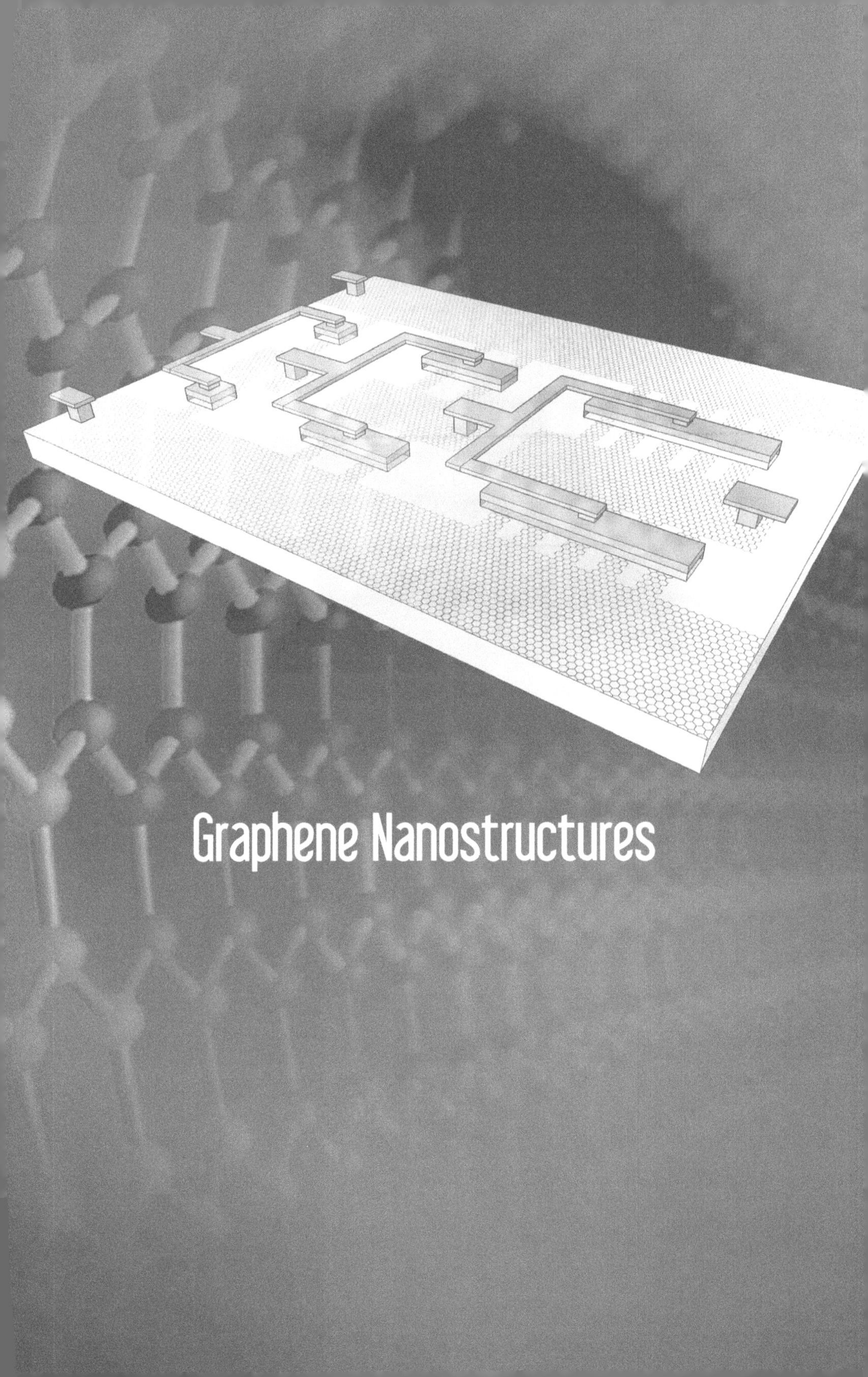

Graphene Nanostructures

Graphene Nanostructures

Modeling, Simulation, and Applications in Electronics and Photonics

Yaser M. Banadaki
Safura Sharifi

JENNY STANFORD
PUBLISHING

Published by

Jenny Stanford Publishing Pte. Ltd.
Level 34, Centennial Tower
3 Temasek Avenue
Singapore 039190

Email: editorial@jennystanford.com
Web: www.jennystanford.com

British Library Cataloguing-in-Publication Data
A catalogue record for this book is available from the British Library.

Graphene Nanostructures: Modeling, Simulation, and Applications in Electronics and Photonics

ISBN 978-981-4800-36-5 (Hardcover)
ISBN 978-0-429-02221-0 (eBook)

Contents

Preface

Over the past few decades, there have been tremendous innovations in electronics and photonics. Integrated circuits are everywhere and an indispensable part of our life, ranging from portable electronics to telecommunications and transportation. It is estimated that more than 1 trillion semiconductor devices shipped for the first time in 2016 with a rising demand due to new application areas such as the Internet of things, wearable electronics, and robotics. According to the Semiconductor Industry Association, the semiconductor industry is responsible for the direct employment of roughly a quarter of a million and indirectly supports another 1 million jobs only in the United States.

The development of ultra-fast-growing technologies mostly relies on fundamental understanding of novel materials with unique properties as well as new designs of device architectures with more diverse and better functionalities. In this regard, the promising approach for next-generation nanoscale electronics and photonics is to exploit the extraordinary characteristics of novel nanomaterials. The discovery of graphene, one atomic layer of carbon sheet, in 2004 has prompted research into its potential in developing future electronic and photonic devices owing to its exceptional electronic, photonic, and mechanical properties.

The driving engine for the exponential growth of digital information processing systems is scaling down the transistor dimensions. For decades, this has enhanced the device performance and density. However, the International Technology Roadmap for Semiconductors (ITRS) states the end of Moore's law in the next decade due to the scaling challenges of silicon-based complementary metal-oxide semiconductor (CMOS) electronics, e.g., extremely high power density. A large group of emerging materials and devices is being extensively studied to replace silicon due to its scaling limit in sight. Germanium has been substituted by silicon roughly half a century ago by moving up on group IV of the periodic table. Interestingly, moving up one more block, we reach carbon, which has been widely tipped as a substitute for next-generation electronics due to its impressive crystal structures, or allotropes. Among carbon allotropes, graphene could be the candidate of future technology as the end of Moore's law approaches. The fabrication technology of graphene is still in the early stages, and engineers need to devise methods for mass production of large, uniform sheets of pure, single-planed graphene, etc. As such, transistor simulation and modeling have been playing an important role for evaluating futuristic graphene-based devices and circuits. In search of non-classical devices and related technologies, graphene-based field-effect transistors are being explored extensively for integrated circuit design. The transistors in integrated circuits are approaching atomic scales. This will soon result in the creation of a growing knowledge gap between the underlying technology and state-of-the-art electronic device modeling and simulations. To bridge this gap, the book provides the computational analysis and mathematical circuit model of a novel transistor with an alternative graphene channel.

Computational analysis plays a vital role in optimizing the transistor structure, and the device models allow designers to input custom design parameters for evaluating the performance of graphene circuits as a promising very large scale integration (VLSI) technology. The primary aim of the book is to inspire and to educate fellow circuit designer and students into the field of emerging low-power and high-performance design based on graphene. As such, the first eight chapters of the book are dedicated to the electronic application of graphene, presenting both computational and analytical models of graphene-based transistors derived from the fundamental understanding of the material properties. For the computational model, a graphene nanoribbon field-effect transistor (GNRFET) has been simulated by solving a quantum transport model based on a self-consistent solution to the three-dimensional (3D) Poisson equation and 1D Schrödinger equations within the non-equilibrium Green's function (NEGF) formalism. The quantum transport model fully treats short-channel-length electrostatic effects and the quantum tunneling effects, leading to the technology exploration of GNRFETs for the future. A comprehensive study of static metrics and switching attributes of GNRFET has been presented, including the performance dependence of device characteristics to the GNR width and the scaling of its channel length down to 2.5 nm. In addition, a physics-based circuit model of GNRFET is presented to be used for integrated circuit design, followed by investigating the effect of edge roughness on the device performance. The accuracy of our developed analytical model has been validated against the device-level atomistic computational model for ideal-edge and rough-edge GNRFETs. The model is also used to study the effect of edge roughness on the power and delay performance of GNRFETs by simulating several graphene-based circuits, including energy recovery logic circuit design. Chapters 2 to 7 provide circuit designers with the recent progress on graphene transistors for nanoscale circuits.

Graphene has been extensively investigated as a promising material for various types of high-performance sensors due to its large surface-to-volume ratio, remarkably high carrier mobility, high carrier density, high thermal conductivity, extremely high mechanical strength, and high signal-to-noise ratio. The power density and the corresponding die temperature can be tremendously high in scaled emerging technology designs, urging the on-chip sensing and controlling of the generated heat in nanometer dimensions. Graphene is also a promising material for energy conversion and storage due to its maximum possible surface-to-bulk ratio along with its exceptional electrical and thermal transports. Graphene can be used for power generation in various applications such as energy harvesting in automotives and integrated circuits, as well as efficient thermal-to-electrical energy conversion in wearable electronics. Graphene is biologically non-invasive, light, flexible, transparent, and mechanically strong, which can convert the generated body heat into an electric current for smart-clothing technology. In Chapter 8, we examine the feasibility of GNRs as temperature sensors and thermoelectric materials. This chapter provides sensor designers with important examples of graphene feasibility for on-chip temperature sensing applications.

There has also been an explosion of interest in graphene for photonic applications as it provides a degree of freedom to manipulate electromagnetic waves. Graphene is a zero-bandgap semimetal that attracted significant attention due to its strong interaction with photons in a wide energy range from microwave to ultraviolet, as well as its high carrier mobility for high-speed applications in a broad wavelength range. The propagation of these waves can be actively controlled by varying the chemical potential in graphene, which can be tuned by chemical doping, voltage bias, external magnetic field, or optical excitation. The absorption coefficient of graphene exceeds $5 \times 10^7\ m^{-1}$ in the visible range if it is normalized to its atomic thickness, which is more than 10 times larger than those in gallium arsenide and silicon. As such, the interaction with light can be further enhanced using graphene multilayers for practical device applications. The last two chapters are dedicated to photonic applications of graphene. We present an aperiodic multilayer structure composed of multiple layers of graphene and hexagonal boron nitride (h-BN) as a selective, tunable, and switchable thermal emitter. Using a genetic optimization algorithm and a transfer matrix code, the layer thicknesses and materials of the aperiodic multilayer structure are optimized to maximize the absorptance/emittance for normal light incidence at a single wavelength.

This book can serve as a textbook for graduate-level course in nanoscale electronics and photonics design. The book can be used by students in electrical and electronics engineering, applied physics, and materials science with a focus on next-generation reduced-dimension materials such as graphene and GNRs. While most of the books focus on either synthesis, fabrication, and characterization or theoretical concepts of graphene, this book shines a light on graphene models and their circuit simulations for the promising applications of graphene in electronics and photonics. This provides a foundation to examine graphene applications in electronics, photonics, and sensor designs. The material covered in the book will be very useful for graduate students carrying out research on graphene-based nanostructures for electronics, photonics, and sensing applications. The book provides a quick review of the fundamental properties and applications of graphene for graduate students to explore how the presented models, designs, and nanostructures can be extended for their thesis and dissertation research. It includes the integration of graphene with sub-nanometer CMOS technology nodes, high-frequency and high-speed electronics, computational and analytical modeling of GNRFETs, design of electrically controllable photonic detectors, sensor electronics, and circuits, and numerous other applications. The content in this book forms the basis to understand the unique nanostructures composed of graphene, which will be very useful for practicing engineers in the field of electronics device design and technology, solid-state physics and semiconductors, nanoscience and nanotechnology, and nanophotonic and nanoelectronic devices and circuits.

The authors would like to thank our advisors Dr. Ashok Srivastava, Dr. Jonathan P. Dowling, and Dr. Georgios Veronis for their comments and sharing immense knowledge. Without their insight and contributions, portions of this book may not have been possible. This book is dedicated to our kids, Persia and Parmida, for their emotional support and patience during the writing of the book. They are just about the best children a parent could hope for: happy, loving, and fun to be with. The book would have not been complete without the encouragement and support of our families.

Yaser M. Banadaki

Safura Sharifi

Spring 2019

Chapter 1

Introduction to Graphene

1.1 Physical Geometry and Properties

Graphene is a one-atom thick sheet of sp^2-hybridized carbon atoms arranged in a honeycomb lattice, and thereby the thinnest possible material in Nature. Since the middle of the 20th century, graphene has been studied theoretically as the building block of carbon allotropes [1]. The hybridized s and p orbitals form strong directional covalent bonds leading to a large number of different allotropes. The nature of these allotropes was first understood by Linus Pauling in his book titled *The Nature of the Chemical Bond* [2], as all have the same basic motif, namely, the benzene ring. The building block of all these allotropes is carbon atoms in two-dimensional (2D) honeycomb lattice structure, called graphene. Hence, graphite can be looked upon as stacked graphene, nanotubes are rolled graphene, and fullerenes are wrapped graphene, as shown in Fig. 1.1 [3]. Fullerenes were discovered in 1985 [4], nanotubes were discovered in 1991 [5], and graphene was discovered in 2004 [6].

Graphene as a material on its own was discovered by Andre Geim and Konstatin Novoselov [7]. However, the history of graphene goes back much further to 1947 when Wallace [8] first described it by calculating the band structure of a single layer of carbon atoms arranged in a hexagonal 2D lattice. The name "graphene" for single carbon layers of the graphitic structure was introduced in 1994 [9],

Graphene Nanostructures: Modeling, Simulation, and Applications in Electronics and Photonics
Yaser M. Banadaki and Safura Sharifi

ISBN 978-981-4800-36-5 (Hardcover), 978-0-429-02221-0 (eBook)
www.jennystanford.com

only 10 years before its discovery. The stability of 2D crystals was not considered to be possible for a long time due to thermodynamic fluctuations which is why graphene's discovery was delayed [10]. The mechanical cleavage of graphite paved the way for isolating a single layer of graphene deposited on an oxidized silicon substrate. This groundbreaking experiment was considered the first observation of a stable 2D material, which led to the award of the Nobel Prize in physics in 2010 [11]. Since its discovery, graphene has soon emerged as a promising nanomaterial for novel applications in electronics, photonics, and sensors due to its remarkable electrical and optical properties, which will be explained to some extent in this chapter.

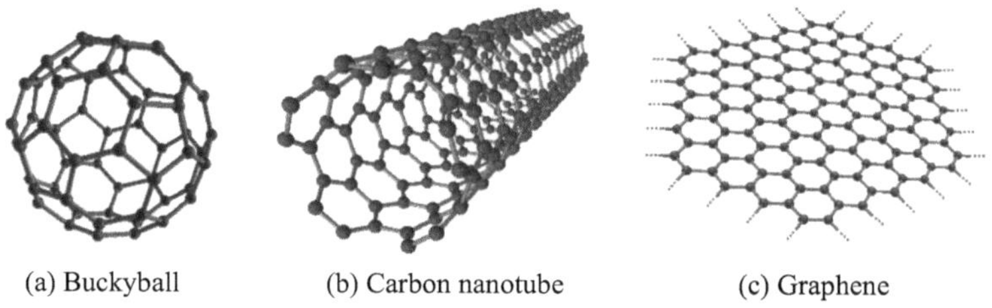

Figure 1.1 Three carbon allotropes: (a) buckyball, (b) carbon nanotube, and (c) graphene [3].

Graphene has a honeycomb-like hexagonal lattice, as shown in Fig. 1.2a. As such, the carbon atoms form strong σ covalent bonds by three in-plane sp^2-hybridized orbitals, whereas the fourth bond is a π bond in z-direction [12]. The electron in this bond can move freely in the delocalized π-electronic system referred as the π-band and π*-bands [13]. The lattice structure of graphene made out of two interpenetrating triangular lattices results in a unit cell consisting of two atoms, as shown in Fig. 1.2b. The lattice vectors can be written as follows:

$$\vec{a_1} = \frac{a_{cc}}{2}\left(3,\sqrt{3}\right),\ \vec{a_2} = \frac{a_{cc}}{2}\left(3,-\sqrt{3}\right) \tag{1.1}$$

where a_{cc} = 1.42 Å is the carbon–carbon distance and (p,q) implies the vector $p\hat{x}+q\hat{y}$, where $\hat{x}$ and $\hat{y}$ are unit vectors along x and y directions. Since electronic transport can be two-dimensional in a graphene lattice, the dispersion relation for graphene has also two dimensions. The reciprocal lattice vectors can be obtained as follows:

$$\overrightarrow{b_1} = \frac{2\pi}{3a_{cc}}\left(1, \sqrt{3}\right),\ \overrightarrow{b_2} = \frac{2\pi}{3a_{cc}}\left(1, -\sqrt{3}\right) \quad (1.2)$$

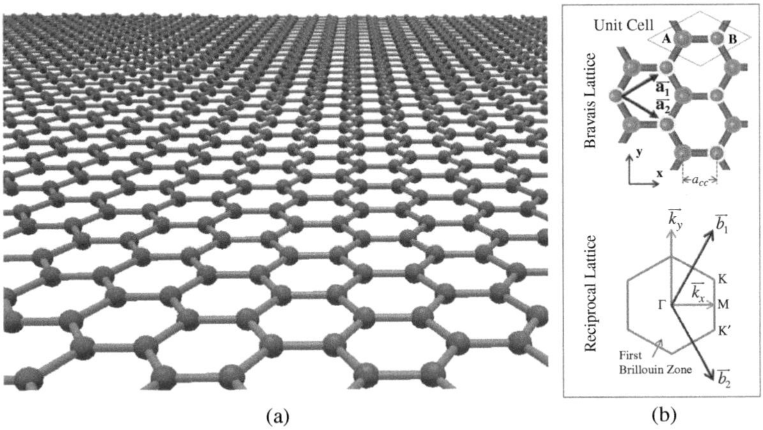

Figure 1.2 (a) 2D honeycomb lattice of graphene, which consists of two triangular sub-lattices. (b) Bravais lattice and reciprocal lattice of graphene [3].

Due to the honeycomb lattice structure, there are two sets of three cone-like points K and K' on the edge of the Brillouin zone named Dirac points, where the conduction and valence bands meet each other in momentum space [12] as follows:

$$K = \left(\frac{2\pi}{3a_{cc}}, \frac{2\pi}{3\sqrt{3}a_{cc}}\right),\ K' = \left(\frac{2\pi}{3a_{cc}}, -\frac{2\pi}{3\sqrt{3}a_{cc}}\right) \quad (1.3)$$

The behavior of charge carriers near Dirac points resembles the Dirac spectrum for massless fermions [14] and can be described by linear dispersion relation as follows:

$$E\left(\overrightarrow{k'}\right) = \pm\hbar\upsilon_F\left|\overrightarrow{k'}\right| \quad (1.4)$$

where k' is the momentum near the Dirac point, $\hbar$ is reduced Planck constant, and υ_F is the Fermi velocity. Charge carriers near Dirac points behave like relativistic particles ideally transporting with Fermi velocity, which is theoretically 300 times smaller than the speed of light [14]. Assuming the first nearest neighbor interaction, the close form of dispersion relation near Dirac points can be obtained [12] as follows:

$$E(\vec{k}) = \pm t\sqrt{1 + 4\cos\frac{\sqrt{3}k_x a_{cc}}{2}\cos\frac{k_y a_{cc}}{2} + 4\cos^2\frac{k_y a_{cc}}{2}} \tag{1.5}$$

where a_{cc} is the carbon–carbon atomic distance, k_x and k_y are wave vectors in x and y directions, and $t = -2.7$ eV is the nearest neighbor hopping energy. Minus and plus signs correspond to the conduction and valence bands, respectively. The graphene band structure has a 2D Brillouin zone in the momentum space with the linear energy dispersion around the band edges instead of quadratic [15], as shown in Figs. 1.3a and 1.3b.

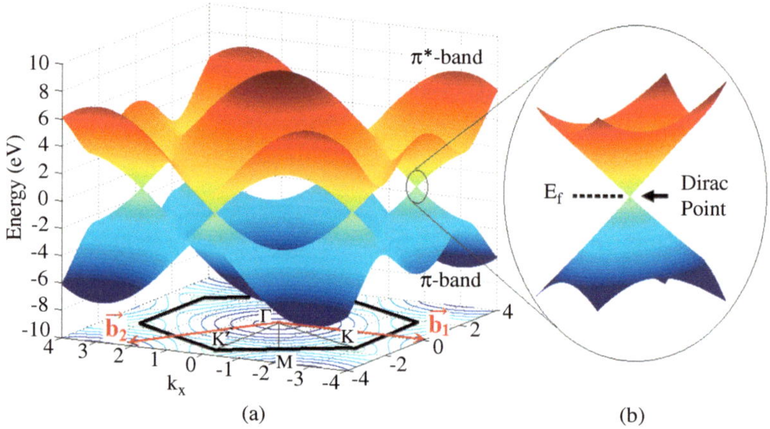

Figure 1.3 (a) Graphene band structure and first Brillouin zone in momentum space. *Note*: The position of Dirac points *K*, *K'* and reciprocal lattice vectors are also shown underneath the graphene band structure. (b) Linear band near Dirac point and the position of Fermi level [3].

Another extraordinary feature of graphene is that there is no energy gap between conduction and valence bands at Dirac points. As such, graphene can be considered a semiconductor with no bandgap, or a semimetal with no band overlap. If no external perturbation and no kinetic energy ($T = 0$ K) are present, the Fermi level E_F (or chemical potential μ_c) is exactly at the touching point between valence and conduction bands. At these points, the electron density is equal to the hole density, corresponding to the minimum value of the total carrier density and the so-called charge neutrality point (CNP). The electric field can modulate the density of states and switch the device from low conductivity states near the Dirac point to high conductivity states elsewhere. Due to the gapless nature of the dispersion, the number of charge carriers can be continuously changed by the electric field effect, adding either electrons or holes to the system. The electrons behave exactly the same as holes due to the symmetric conduction and valence band structures.

Graphene shows perfect ambipolar electric field effect so that its charge carrier parity can be tuned to be either n or p type by applying electric field with different polarity. Interestingly, it is possible to achieve different carrier parity by varying the DC bias voltage, i.e., the perpendicular electric field, unlike traditional silicon technology wherein the materials need to be actually doped by other materials. At the CNP, there is still a finite amount of current flowing through the low conduction state near the Dirac point, resulting in a non-negligible dark current for transistor applications. Furthermore, the charge density never vanishes completely due to the existence of electron–hole puddles on the SiO_2 substrate, leading to minimum conductance on the order of $\sim e^2/h$ [16].

The carrier transport in graphene is similar to the transport of massless particles since 2D electron gas in graphene [17] provides both high carrier velocity and high carrier concentration, resulting in large carrier mobility and consequently its faster switching capability [14]. While the bottleneck of scaling silicon channel is in heat removal of dissipated power, graphene has excellent thermal conductivity due to strong carbon–carbon bonding [18]. Atomically thin structure of monolayer graphene results in better gate control over the channel, and the planar structure is compatible with current CMOS fabrication processes introducing the potential production of wafer-scale integrated circuits [19, 20]. Graphene and related 2D materials could be utilized in heterostructures to create light-emitting devices for the next generation of thin, flexible, and transparent electronics [21, 22]. Graphene shows some interesting properties in sensing applications as its planar geometry of graphene with one carbon atom thickness maximizes the active sensing area [23]. As large-area graphene is bendable and printable, the deposition of graphene on flexible substrates opens the door to high-frequency low-voltage flexible applications [24]. Scanning tunneling microscopy (STM) measurements show that local charge fluctuations of graphene on flat hBN are much smaller than on SiO_2 substrate [25]. Electron transport in graphene has low scattering rate due to intrinsic phonons, and thus the mobility at room temperature can reach above 100,000 $cm^2V^{-1}s^{-1}$ on a device encapsulated between two layers of hBN [26]. This value is far beyond the mobility of graphene on SiO_2, which is on the order of a few thousands $cm^2V^{-1}s^{-1}$ [27].

1.2 Graphene Nanoribbon

Graphene shows exotic electronic properties and may outperform state-of-the-art silicon in many applications [28, 29] due to its exceptional properties such as large carrier motility, high carrier concentration, high thermal conductivity, and atomically thin planar structure [14, 18]. However, the application of large-area graphene is limited for integrated circuits due to the lack of bandgap and the need for only narrow stripes of graphene. The latter are known as graphene nanoribbons (GNRs), which are promising alternative as replacement of transistor channels [30, 31] for next-generation integrated circuits. In principle, the GNRs can be produced by patterning large-area graphene using more standard fabrication methods with much more controllability than CNTs, whose chiralities are statistically predetermined during the manufacture process. While CNTs require a different set of processing techniques, the younger counterpart, graphene, shares a similar set of processing techniques currently used for silicon. GNR can be fabricated from large-area graphene using high-resolution lithography like e-beam lithography. GNRs share many of the fascinating electrical [32], mechanical [33], and thermal [18, 34] properties of CNTs such as large carrier mobility and thermal conductivity [35]. The mean free path (MFP) of electron in GNRs with smooth edge is comparable with CNT and can reach to micrometer range [36]. In addition, GNR has a very large current conduction capacity (1000 times larger than Cu) with extraordinary mechanical strength and thermal conductivity.

Width confinement of graphene down to the sub-10 nm scale is essential to open a bandgap that is sufficient for room temperature transistor operation. The size of the induced energy gap is a direct function of the nanoribbon width, such that decreasing the width increases the bandgap [37, 38]. For exfoliated GNR with a width of 15 nm, the bandgap of 0.3 eV was first measured in 2007 [39]. The induced bandgap in excess of 1 eV can be opened for a GNR with a width below 2 nm [40]. Several experimental methods have been already proposed for narrowing the width by etching down GNR to 4 nm [41] and chemical synthesis down to 2 nm [42]. Other lithography methods based on atomic force microscopy (AFM) [43] and STM [44] have been proposed for the fabrication of GNRs. Graphene nanoribbons with a few nanometer width can be produced by unzipping carbon nanotubes [45] with bottom-up chemical approach [46]. This method can reduce the edge roughness induced by e-beam lithography and recover zigzag or armchair edges of GNRs [47]. Mass production of GNRs can be made possible by using multi-walled CNTs (MWCNTs) as precursors such that the GNR widths can be controlled by controlling the size of the starting MWCNTs and the conditions of dry etching [48] or solution-based oxidative process [49].

The ribbon width is not the only factor, and the nano-cutting of large-area graphene needs special attention on the type of edge boundary (or chiral angle) as it can determine whether the GNRs are metallic or semiconducting. The chiral angle represents the crystallographic direction of the axis of the GNR and comes from theoretical studies of CNTs corresponding to chiral vector $\vec{C}_{\mathrm{h}} = n\vec{a}_1 + m\vec{a}_2$, and CNT indices: CNT($n$,$m$), where $\vec{a}_1$ and $\vec{a}_2$ are the unit vectors for the graphene hexagonal structure, n and m are the integer coefficients along the $\vec{a}_1$ and $\vec{a}_2$ directions, as shown in Fig. 1.4a. This unique notation of an individual CNT traces the CNT around its circumference from one carbon atom (called the reference point) back to itself. The atomistic structure of GNRs can be considered as the unfolded of the corresponding CNTs with the desired width to adopt the same terminology for the GNRs. Among all the possible chiral angles (or CNT indices), special attention is paid to the zigzag GNRs(n, n), and armchair GNRs(n, 0). For example, the circumference edge of CNT(n,n) is along the armchair direction and thereby unzipping this CNT results in the zigzag-edge GNR(n,n). Figure 1.4b shows the different nano-cutting directions of graphene lattice for producing nanoribbons with armchair and zigzag edges. The angles between zigzag and armchair edges are multiples of 30 degrees, such that GNRs with either zigzag or armchair edges can be chosen by changing the direction of nano-cutting by 30 degrees.

The optical and AFM images of the graphene sheet are shown in Fig. 1.5a [50]. In the images, the crystalline orientations of the graphene sheet as well as zigzag and armchair edges can be identified.

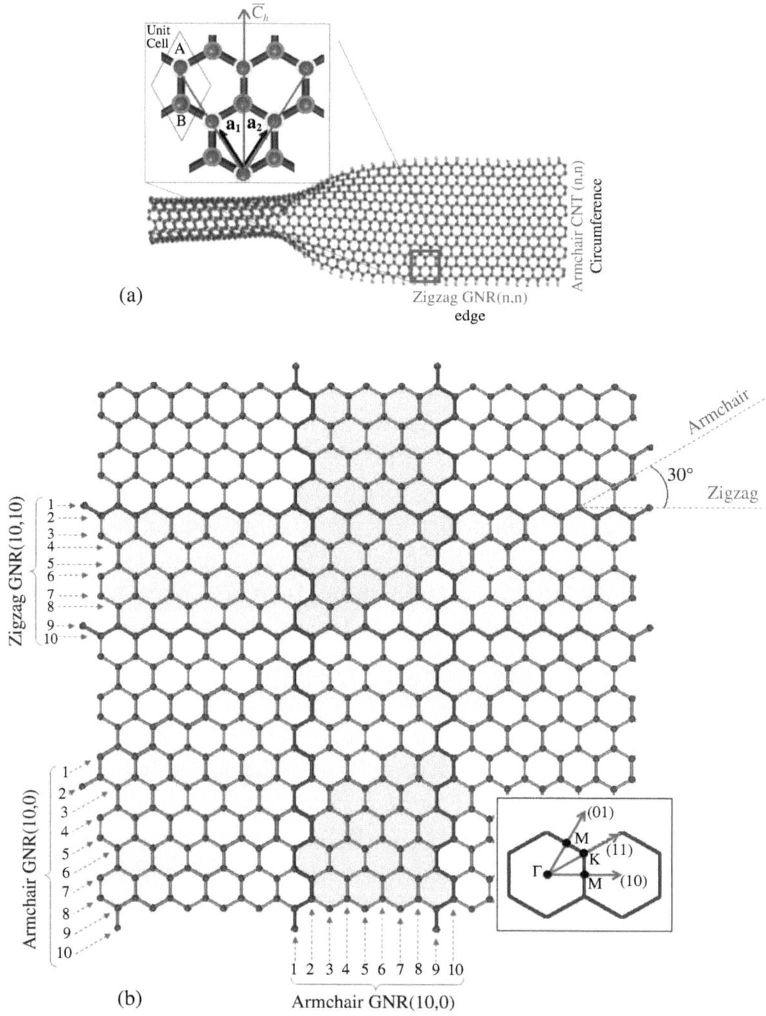

Figure 1.4 (a) Unfolding carbon nanotube with armchair edge results in graphene nanoribbon with zigzag edge. (b) Cuts along two directions of a graphene sheet to produce zigzag (red) and armchair (green, blue) termination of the GNRs [3].

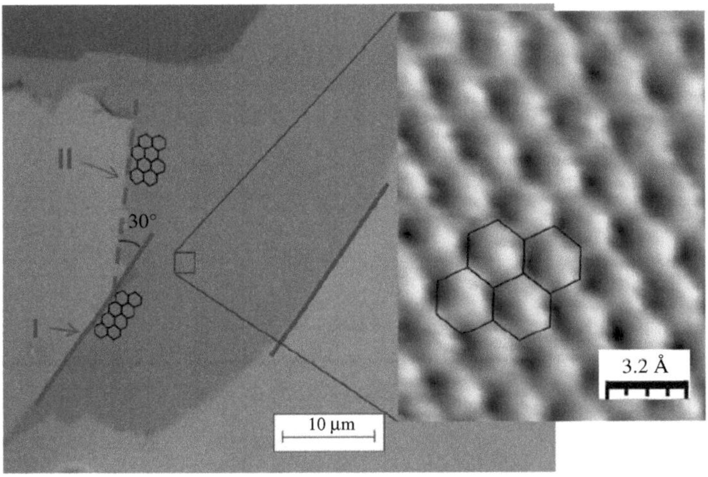

(a)

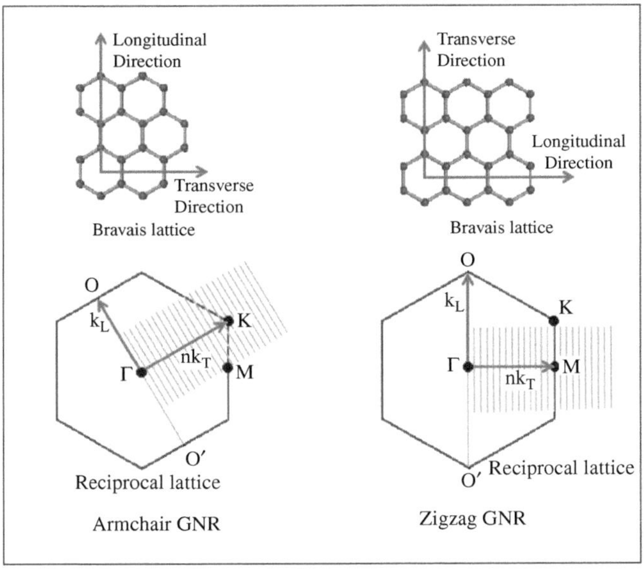

(b)

Figure 1.5 (a) Optical image of the graphene sheet (left) and a lattice resolution AFM image (right) to identify the graphene crystalline orientation. Superimposing the hexagons onto the optical image, the crystallographic orientations of edges I (zigzag) and II (armchair) are shown. Reprinted from Ref. [50], Copyright 2011, with permission from AIP Publishing. (b) Discretized transverse wavevector of armchair and zigzag graphene nanoribbons due to the confinement in transverse direction [3].

The electronic structure of a GNR can be obtained from that of infinite graphene. Zigzag and armchair GNRs can be produced from an infinite graphene sheet by cutting in the (10) and (11) directions, respectively, in a 2D space. In the reciprocal space, these directions correspond to the *Γ*-*M* and *Γ*-*K* paths in the Brillion zone for zigzag and armchair terminations, respectively. The wavevector in the transverse direction, k_T, becomes quantized, whereas the longitudinal wavevector k_L remains continuous for a GNR of infinite length, as shown in Fig. 1.5b. Thus, the energy bands consist of a set of 1D energy dispersion relations, which are cross sections of those for infinite graphene. The energy dispersion relations of 2D graphene are shifted from OO′ by discretized reciprocal vector nk_T (n = 1, 2,..., N − 1) in parallel with k_L, resulting in N pairs of 1D energy dispersion curves corresponding to the cross sections of the 2D energy dispersion surface. If the cutting line passes through the Dirac point of the 2D Brillouin zone, where the conduction and the valence energy bands of pristine graphene touch each other, the 1D energy spectra have a zero energy gap [51]. This corresponds to zigzag GNRs with conducting behavior as shown in Fig. 1.6a,b [3].

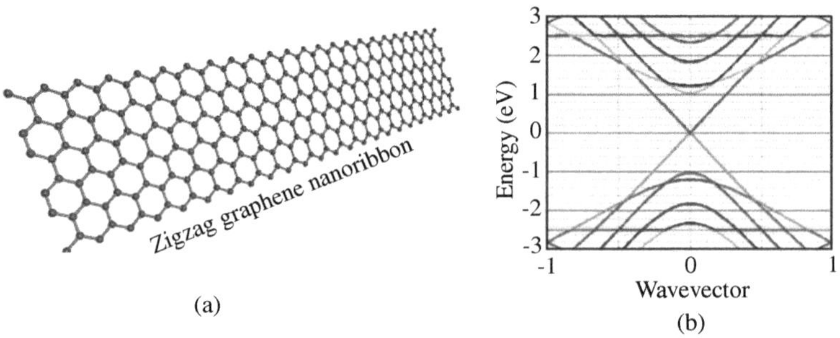

Figure 1.6 (a) Schematic of a zigzag GNR. (b) The energy dispersion graphs for the zigzag GNR in (a) [3].

The armchair GNR can yield either conducting or semiconducting characteristics depending on the number of atoms in transverse direction. The electronic structure of armchair GNRs is closely related to that of zigzag CNTs and needs to be classified into three groups as their bandgap changes with a period-three modulation depending on the number of atoms in confined transverse direction. For an arbitrary integer p, two thirds of armchair GNRs, (3p,0) and

(3*p*+1,0), are semiconducting while the third subclass, (3*p*+2,0), has a very small bandgap showing metallic behavior. The first-principle calculation can be used to obtain the electronic structure of graphene nanoribbon. It can be solved either by Dirac's equation of massless particles with an effective speed of light [52] or simple tight-binding approximation [37, 53]. Figure 1.7a shows the bandgap of each GNR group versus the number of dimer lines in transverse direction. Figure 1.7b shows the calculated dispersion relations of

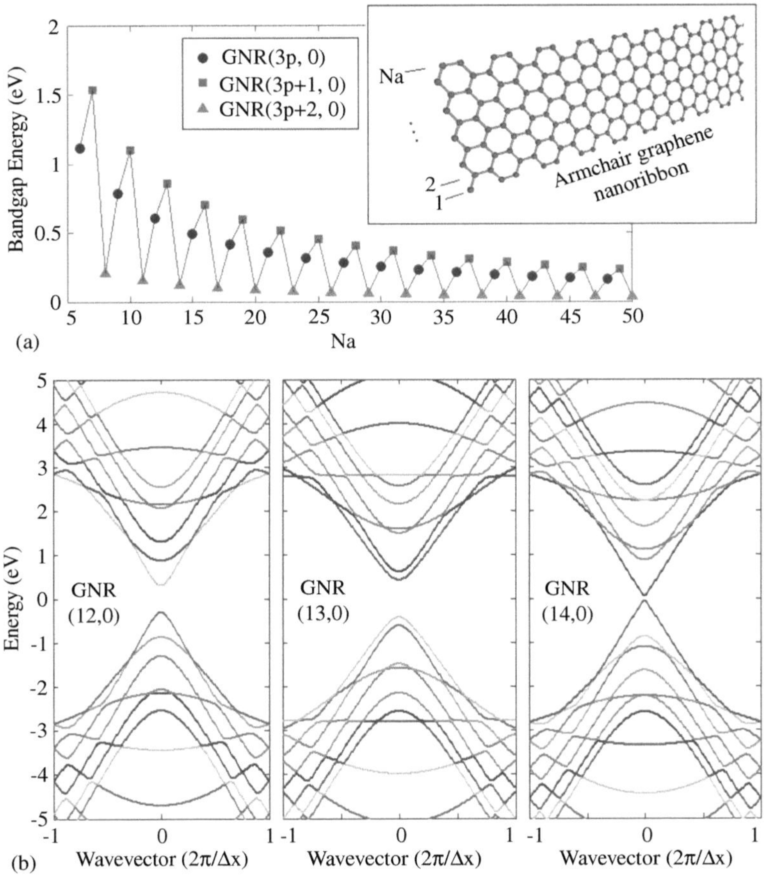

Figure 1.7 (a) Bandgap energy of three GNR families of armchair GNR versus GNR index. *p* is an arbitrary integer larger than 2. The inset shows the schematic of an armchair GNR and the description of number of dimer lines. (b) Energy dispersion relation of GNR(12,0), GNR(13,0), and GNR(14,0) [3].

GNR(12,0), GNR(13,0), and GNR(14,0) as a representative of three GNR families (N,0) = (3*p*,0), (3*p*+1,0), and (3*p*+2,0), respectively, where *p* is an arbitrary integer. Removing or adding one edge atom along the nanoribbon can significantly change the bandgap energy of the GNR.

References

1. Geim, A. K. and Novoselov, K. S. (2007). The rise of graphene, *Nature Materials*, **6**, pp. 183–191.
2. Pauling, L. (1960). *The Nature of the Chemical Bond*, vol. 3., Cornell University Press, Ithaca, NY.
3. Mohammadi Banadaki, Y. (2016). Physical modeling of graphene nanoribbon field effect transistor using non-equilibrium Green function approach for integrated circuit design, *LSU Doctoral Dissertations* 1052, https://digitalcommons.lsu.edu/gradschool_dissertations/1052
4. Kroto, H. W., Heath, J. R., O'Brien, S. C., Curl, R. F., and Smalley, R. E. (1985). C 60: Buckminsterfullerene, *Nature*, **318**, pp. 162–163.
5. Iijima, S. (1991). Helical microtubules of graphitic carbon, *Nature*, **354**, pp. 56–58.
6. Van Noorden, R. (2006). Moving towards a graphene world, *Nature*, **442**, pp. 228–229.
7. Novoselov, K. S., Geim, A. K., Morozov, S. V., Jiang, D., Zhang, Y., Dubonos, S. V., Grigorieva, I. V., and Firsov, A. A. (2004). Electric field effect in atomically thin carbon films, *Science*, **306**, pp. 666–669.
8. Wallace, P. R. (1947). The band theory of graphite, *Physical Review*, **71**, p. 622.
9. Boehm, H. P., Setton, R., and Stumpp, E. (1994). Nomenclature and terminology of graphitic intercalation compounds, *Pure Appl. Chem.*, **66**, pp. 1893–1901.
10. Mermin, N. D. (1968). Crystalline order in two dimensions, *Physical Review*, **176**, p. 250.
11. Foundation, T. N. The Nobel Prize in Physics 2010, *http://www.nobelprize.org/nobelprizes/physics/laureates/2010/.*
12. Neto, A. C., Guinea, F., Peres, N., Novoselov, K. S., and Geim, A. K. (2009). The electronic properties of graphene, *Reviews of Modern Physics*, **81**, p. 109.

13. Stampfer, C., Fringes, S., Güttinger, J., Molitor, F., Volk, C., Terrés, B., Dauber, J., Engels, S., Schnez, S., and Jacobsen, A. (2011). Transport in graphene nanostructures, *Frontiers of Physics*, **6**, pp. 271–293.

14. Cooper, D. R., D'Anjou, B., Ghattamaneni, N., Harack, B., Hilke, M., Horth, A., Majlis, N., Massicotte, M., Vandsburger, L., and Whiteway, E. (2012). Experimental review of graphene, *ISRN Condensed Matter Physics*, **2012**, 501686, p. 56.

15. Avouris, P. (2010). Graphene: Electronic and photonic properties and devices, *Nano Letters*, **10**, pp. 4285–4294.

16. Tan, Y.-W., Zhang, Y., Stormer, H. L., and Kim, P. (2007). Temperature dependent electron transport in graphene, *The European Physical Journal-Special Topics*, **148**, pp. 15–18.

17. Novoselov, K., Geim, A. K., Morozov, S., Jiang, D., Katsnelson, M., Grigorieva, I., Dubonos, S., and Firsov, A. (2005). Two-dimensional gas of massless Dirac fermions in graphene, *Nature*, **438**, pp. 197–200.

18. Banadaki, Y., Mohsin, K., and Srivastava, A. (2014). A graphene field effect transistor for high temperature sensing applications, in: *Proc. SPIE 9060, Nanosensors, Biosensors, and Info-Tech Sensors and Systems 2014*, V. K. Varadan (Ed.), vol. 9060, International Society for Optics and Photonics.

19. Li, X., Cai, W., An, J., Kim, S., Nah, J., Yang, D., Piner, R., Velamakanni, A., Jung, I., and Tutuc, E. (2009). Large-area synthesis of high-quality and uniform graphene films on copper foils, *Science*, **324**, pp. 1312–1314.

20. Todri-Sanial, A., Dijon, J., and Maffucci, A. (Eds.) (2017). *Carbon Nanotubes for Interconnects: Process, Design and Applications*, Springer International Publishing.

21. Eda, G., Fanchini, G., and Chhowalla, M. (2008). Large-area ultrathin films of reduced graphene oxide as a transparent and flexible electronic material, *Nature Nanotechnology*, **3**, pp. 270–274.

22. Kim, P., Han, M. Y., Young, A. F., Meric, I., and Shepard, K. L. (2009). Graphene nanoribbon devices and quantum heterojunction devices, in: *2009 IEEE International Electron Devices Meeting (IEDM)*, December 7–9, 2009, Baltimore, MD, USA, IEEE.

23. Hill, E. W., Vijayaragahvan, A., and Novoselov, K. (2011). Graphene sensors, *IEEE Sensors Journal*, **11**, pp. 3161–3170.

24. Petrone, N., Meric, I., Chari, T., Shepard, K. L., and Hone, J. (2015). Graphene field-effect transistors for radio-frequency flexible electronics, *IEEE Journal of the Electron Devices Society*, **3**, pp. 44–48.

25. Xue, J., Sanchez-Yamagishi, J., Bulmash, D., Jacquod, P., Deshpande, A., Watanabe, K., Taniguchi, T., Jarillo-Herrero, P., and LeRoy, B. J. (2011). Scanning tunnelling microscopy and spectroscopy of ultra-flat graphene on hexagonal boron nitride, *Nature Materials*, **10**, pp. 282–285.

26. Wang, L., Meric, I., Huang, P., Gao, Q., Gao, Y., Tran, H., Taniguchi, T., Watanabe, K., Campos, L., and Muller, D. (2013). One-dimensional electrical contact to a two-dimensional material, *Science*, **342**, pp. 614–617.

27. Chen, J.-H., Jang, C., Xiao, S., Ishigami, M., and Fuhrer, M. S. (2008). Intrinsic and extrinsic performance limits of graphene devices on SiO_2, *Nature Nanotechnology*, **3**, pp. 206–209.

28. Novoselov, K. S., Fal, V., Colombo, L., Gellert, P., Schwab, M., and Kim, K. (2012). A roadmap for graphene, *Nature*, **490**, pp. 192–200.

29. Obeng, Y., De Gendt, S., Srinivasan, P., Misra, D., Iwai, H., Karim, Z., Hess, D., and Grebel, H. (Eds.) (2009). *Graphene and Emerging Materials for Post-CMOS Applications*, The Electrochemical Society.

30. Banadaki, Y. and Srivastava, A. (2015). Scaling effects on static metrics and switching attributes of graphene nanoribbon FET for emerging technology, *IEEE Transactions on Emerging Topics in Computing*, doi:10.1109/TETC.2015.2445104.

31. Obradovic, B., Kotlyar, R., Heinz, F., Matagne, P., Rakshit, T., Giles, M., Stettler, M., and Nikonov, D. (2006). Analysis of graphene nanoribbons as a channel material for field-effect transistors, *Applied Physics Letters*, **88**, p. 142102.

32. Meric, I., Han, M. Y., Young, A. F., Ozyilmaz, B., Kim, P., and Shepard, K. L. (2008). Current saturation in zero-bandgap, top-gated graphene field-effect transistors, *Nature Nanotechnology*, **3**, pp. 654–659.

33. Lee, C., Wei, X., Kysar, J. W., and Hone, J. (2008). Measurement of the elastic properties and intrinsic strength of monolayer graphene, *Science*, **321**, pp. 385–388.

34. Balandin, A. A., Ghosh, S., Bao, W., Calizo, I., Teweldebrhan, D., Miao, F., and Lau, C. N. (2008). Superior thermal conductivity of single-layer graphene, *Nano Letters*, **8**, pp. 902–907.

35. Sarma, S. D., Adam, S., Hwang, E., and Rossi, E. (2011). Electronic transport in two-dimensional graphene, *Reviews of Modern Physics*, **83**, p. 407.

36. Berger, C., Song, Z., Li, X., Wu, X., Brown, N., Naud, C., Mayou, D., Li, T., Hass, J., and Marchenkov, A. N. (2006). Electronic confinement and

coherence in patterned epitaxial graphene, *Science*, **312**, pp. 1191–1196.

37. Son, Y.-W., Cohen, M. L., and Louie, S. G. (2006). Energy gaps in graphene nanoribbons, *Physical Review Letters*, **97**, p. 216803.
38. Barone, V., Hod, O., and Scuseria, G. E. (2006). Electronic structure and stability of semiconducting graphene nanoribbons, *Nano Letters*, **6**, pp. 2748–2754.
39. Han, M. Y., Özyilmaz, B., Zhang, Y., and Kim, P. (2007). Energy band-gap engineering of graphene nanoribbons, *Physical Review Letters*, **98**, p. 206805.
40. Yang, L., Park, C.-H., Son, Y.-W., Cohen, M. L., and Louie, S. G. (2007). Quasiparticle energies and band gaps in graphene nanoribbons, *Physical Review Letters*, **99**, p. 186801.
41. Wang, X., and Dai, H. (2010). Etching and narrowing of graphene from the edges, *Nature Chemistry*, **2**, pp. 661–665.
42. Li, X., Wang, X., Zhang, L., Lee, S., and Dai, H. (2008). Chemically derived, ultrasmooth graphene nanoribbon semiconductors, *Science*, **319**, pp. 1229–1232.
43. Lu, G., Zhou, X., Li, H., Yin, Z., Li, B., Huang, L., Boey, F., and Zhang, H. (2010). Nanolithography of single-layer graphene oxide films by atomic force microscopy, *Langmuir*, **26**, pp. 6164–6166.
44. Tapasztó, L., Dobrik, G., Lambin, P., and Biró, L. P. (2008). Tailoring the atomic structure of graphene nanoribbons by scanning tunnelling microscope lithography, *Nature Nanotechnology*, **3**, pp. 397–401.
45. Srivastava, A., Liu, X., and Banadaki, Y. (2017). Overview of carbon nanotube interconnects, in: *Carbon Nanotubes for Interconnects*, Springer, pp. 37–80.
46. Xie, L., Wang, H., Jin, C., Wang, X., Jiao, L., Suenaga, K., and Dai, H. (2011). Graphene nanoribbons from unzipped carbon nanotubes: Atomic structures, Raman spectroscopy, and electrical properties, *Journal of the American Chemical Society*, **133**, pp. 10394–10397.
47. Kosynkin, D. V., Higginbotham, A. L., Sinitskii, A., Lomeda, J. R., Dimiev, A., Price, B. K., and Tour, J. M. (2009). Longitudinal unzipping of carbon nanotubes to form graphene nanoribbons, *Nature*, **458**, pp. 872–876.
48. Jiao, L., Zhang, L., Wang, X., Diankov, G., and Dai, H. (2009). Narrow graphene nanoribbons from carbon nanotubes, *Nature*, **458**, pp. 877–880.

49. Zhang, Z., Sun, Z., Yao, J., Kosynkin, D. V., and Tour, J. M. (2009). Transforming carbon nanotube devices into nanoribbon devices, *Journal of the American Chemical Society*, **131**, pp. 13460–13463.

50. Almeida, C., Carozo, V., Prioli, R., and Achete, C. (2011). Identification of graphene crystallographic orientation by atomic force microscopy, *Journal of Applied Physics*, **110**, p. 086101.

51. Chen, Z., Lin, Y.-M., Rooks, M. J., and Avouris, P. (2007). Graphene nano-ribbon electronics, *Physica E: Low-dimensional Systems and Nanostructures*, **40**, pp. 228–232.

52. Brey, L. and Fertig, H. (2006). Electronic states of graphene nanoribbons studied with the Dirac equation, *Physical Review B*, **73**, p. 235411.

53. Saito, R., Dresselhaus, G., and Dresselhaus, M. S. (1998). *Physical Properties of Carbon Nanotubes*, World Scientific Publishing Company.

Chapter 2

Graphene for Integrated Circuits

2.1 Introduction

Graphene, a single atomic layer of graphite, has gained remarkable research interest in the past decade by breaking so many records in terms of electrical and thermal conductions as well as mechanical strength. Graphene is a zero-overlap semimetal with very high electrical conductivity due to one highly mobile electron above and below the graphene sheet. The linear dispersion relation at Dirac points corresponds to the existence of electrons and holes with zero effective mass. However, the density of states is zero at the Dirac points and thus the Fermi level must be sifted to higher energies by doping the graphene sheet to avoid low electronic conductivity at ON-state. In addition, the bandgap must be opened by patterning graphene in the form of nanoribbons to avoid high leakage current at OFF-state. Due to the scaling challenge of silicon electronics, the research on graphene nanoribbon (GNR) is in progress to replace silicon with graphene channel in field-effect transistors (FETs). In this chapter, we will introduce the promising applications of graphene in electronics and photonics.

Graphene Nanostructures: Modeling, Simulation, and Applications in Electronics and Photonics
Yaser M. Banadaki and Safura Sharifi

ISBN 978-981-4800-36-5 (Hardcover), 978-0-429-02221-0 (eBook)
www.jennystanford.com

2.2 Scaling Challenges of Silicon Electronics

The evolution of integrated circuits has been largely governed by Moore's law, which was postulated in 1965 [1] by Gordon Moore, cofounder of Intel Corporation. Moore's law states that the number of transistors on a single chip doubles approximately every 18 months. The exponential trend in scaling silicon transistors has enhanced the device performance and density, satisfying the prediction of Moore's law for decades. Scaling down in each new generation has approximately doubled logic circuit density, reduced cost-per-function, and increased the performance and the memory capacity by four times.

This evolution has been mainly enabled by continuous progress in silicon-based complementary metal-oxide semiconductor (CMOS) technology. The scalability of CMOS comes with other unique device properties, including high input resistance, negligible static power dissipation, higher performance by about 40% as well as the simple device fabrication such as self-isolation, simple layout and process steps. However, many factors need to be taken under consideration with continued CMOS scaling, which present challenges for the future and, ultimately, fundamental limits.

The historical growth cannot be maintained only by the conventional scaling theory. In sub-10 nm channel length, the drain–source leakage current significantly increases due to short channel effects. The leakage current is contributed from reverse-biased p–n junction current, weak inversion, and drain-induced barrier lowering (DIBL) [2]. Increased power density and the corresponding dissipated heat in nanometer dimension have also imposed several fundamental physical challenges for silicon [3, 4], seriously affecting the performance of the chip.

Scaling of MOS structure can be divided into three intervals as shown in Fig. 2.1 [2]. While pure lithography could accomplish the task of scaling until 2002, scaling alone by advancing the lithography technology is not sufficient and innovation was required since then. Fabricating more complex device geometries, e.g., multi-gate or nanowire transistor structure, for higher drive current at the constant over-drive voltage and increasing equivalent oxide thickness (EOT) scaling with high-k/metal gate stack were the natural evolution to enhance the electrostatic control over the channel and consequently increase the device robustness to short channel effects. However, more forward-looking solutions for scaling challenges of silicon electronics are improving carrier mobility by utilizing alternate channel materials such that they can be likely solved by the genesis of new materials for integrated circuit [2].

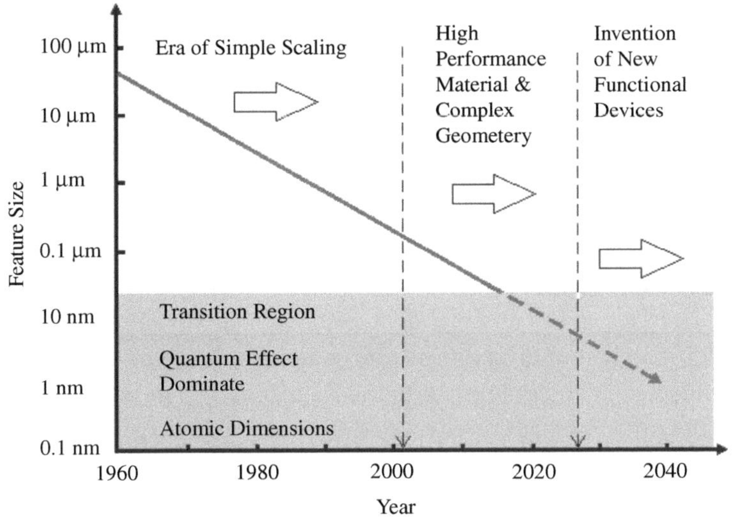

Figure 2.1 Integrated circuit scaling history and projection.

A large group of emerging materials and devices is being extensively studied to replace silicon due to its scaling limit in sight [3]. Germanium has been substituted by silicon roughly half a century ago by moving up on the group IV of periodic table. Interestingly, moving up one more block, we reach to carbon, which has been widely tipped as substitute for next-generation electronics due to its impressive crystal structures, or allotropes. Although silicon and carbon have similar chemical properties due to the same number of electrons in the outermost electronic shell, they have different Coulomb interactions and consequently different size of the electronic wave functions. Thus, the corresponding energies of respective electron systems vary significantly leading to different electronic behavior in most carbon allotropes.

2.3 Graphene-Based Field-Effect Transistors

An FET consists of four terminals: gate, source, drain, and substrate together with insulating dielectric over a conducting channel, as shown in Fig. 2.2a. The Fermi energy of carriers in the channel will rise in the presence of an applied electric field corresponding to the applied gate voltage and needs to be placed in the middle of bandgap to turn the device off because it can minimize both the electrons in conduction band and holes in valence band and thereby minimizes the contribution of electrons and holes in leakage current [5]. Figure 2.2b shows the three Fermi levels in correspondence with three gate voltages applied to a graphene with zero bandgap. At positive V_{GS1}, the Fermi level (E_{F1}) is near or inside the conduction band and electrons contribute to current transport. Decreasing the gate voltage shifts the Fermi level toward the valence band; the total carrier in the channel decreases and consequently minimizes at V_{GS2} corresponding to the charge neutrality point (CNP), where the electron density is equal to the hole density. The equal densities of electrons and holes correspond to the equal contributions of electrons and holes to the total drain–source current due to the same effective mass of conduction and valence bands. The populations of carriers in conduction and valence bands follow the Fermi–Dirac distribution function and can be significant due to the lack of bandgap, and thereby the transistor cannot be fully off by placing the Fermi level in the middle of conduction and valence bands. While this is not an issue for analog applications and graphene has still potential due to very high mobility [6], this limits its application as logic transistors [7, 8].

The semimetallic nature of graphene with overlapping bandgap is clearly an obstacle with regards to its application in semiconducting devices as it cannot be fully switched off by tuning the Fermi level at the energy that conduction and valence bands touch each other [9]. Most gated graphene FETs on various substrates showed I_{ON}/I_{OFF} ratios less than 50, while it needs to be between 10^4 and 10^7 to compete with what is currently required in traditional silicon MOSFETs [3].

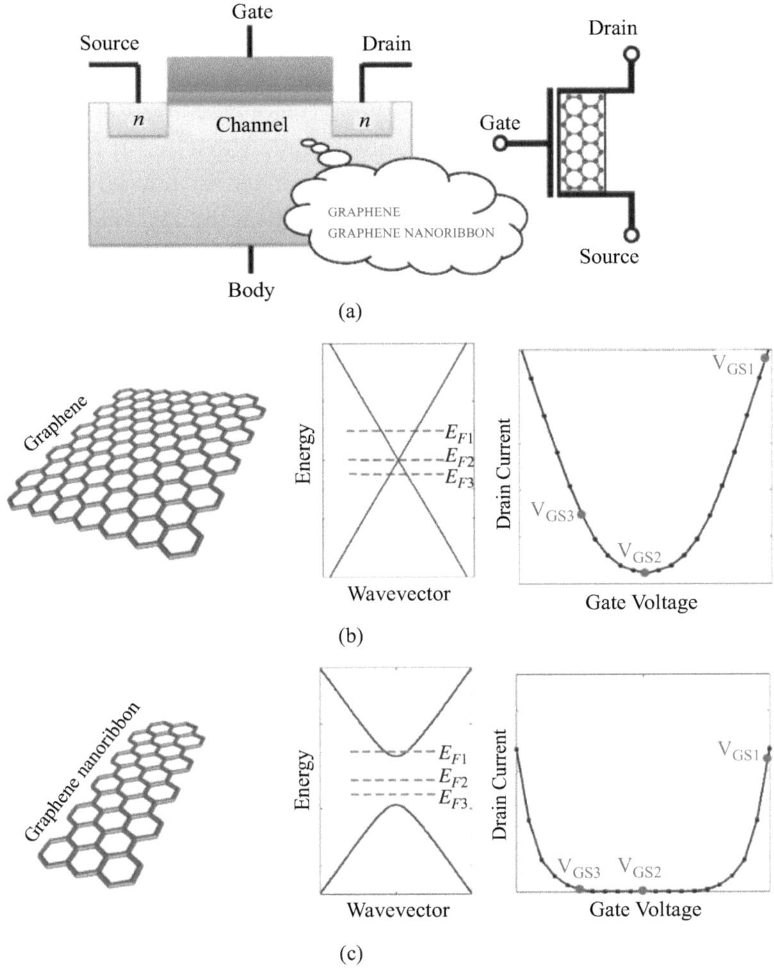

Figure 2.2 (a) Schematic of a field-effect transistor (FET) and symbol of graphene FET. (b) Large-area graphene with zero bandgap. Three Fermi levels are shown in *E–k* diagram and the corresponding gate voltages are also shown in its current–voltage characteristic. (c) Graphene nanoribbon with opened bandgap. Similarly, three Fermi levels are shown in the *E–k* diagram and the corresponding gate voltages are also shown in its current–voltage characteristic.[1]

In order to turn off an FET device with graphene channel, a bandgap of several hundred meV is required, and thus opening

[1]Two graphs in (b) and (c) are sketched to convey the concept, i.e., the importance of bandgap in transfer characteristics, and are not in actual, exact scale with each other.

the bandgap is the most important task in making the graphene transistor become a practical channel material. Patterning large-area graphene into nanoribbon strips can split up two-dimensional (2D) energy dispersion into multiple 1D modes due to the quantum confinement of carriers in 1D graphene, called GNR [8, 10]. Producing GNRs as a way to induce a bandgap is widely considered to be the most elegant and useful methodology due to the fact that keeping device dimensions at the nanoscale dimension urged by the scaling trend of silicon as well. As can be seen in Fig. 2.2c, Fermi level can be placed in the middle of bandgap where the total number of electrons in the conduction band and holes in the valence band are minimized leading to a very small leakage current. A GNRFET can be used in logic circuits in much the same way as in CMOS logic [11].

2.4 Graphene-Based Integrated Circuits

After the discovery of graphene in 2004 [12], it was soon considered one of the promising alternative materials for electronic applications. However, the application of large-area graphene is limited for integrated circuits, and GNR is a promising alternative as a replacement of transistor channel [13, 14] and interconnect [15, 16] for next-generation very large scale integration (VLSI) circuits. Figure 2.3a shows the 3D view of an all-graphene circuit for an example of inverter chains together with its circuit implementation. In this structure, the devices and interconnects are concurrently fabricated by monolithically patterning a single sheet of graphene. This would bring some release from the contact resistance of metal-to-graphene contacts as the material for producing both devices and interconnects is graphene [17, 18]. Pattering GNR with larger width and along zigzag-edge orientation results in GNR interconnects with very small bandgap (Fig. 2.3b) [19]. For using graphene as channel material and local interconnects, both the width and the edge type of GNRs are critically important. This can potentially reduce the complex fabrication process for local interconnect in nanoscale dimensions, leading to ultra-dense and thin integrated circuits [20]. While a modern-day CMOS circuit has approximately 10 interconnect layers, this structure can reduce the number of intra-layer local interconnects for gate-level design. However, it is not possible to get rid of metal contacts and interconnects completely since the gate and source/drain electrodes cannot share the same graphene sheet. The contact resistances of metal–graphene junctions are reported in the range of 1 KΩ to 100 KΩ, with a nominal value of R_{via} = 20 KΩ for via connections with 50 nm width [21].

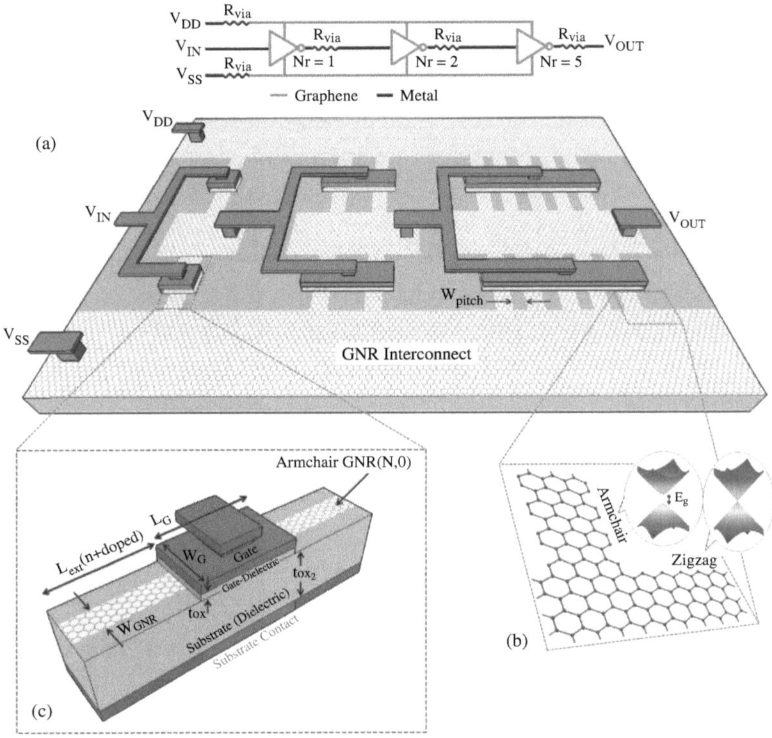

Figure 2.3 (a) 3D view of all-graphene circuit for an example of inverter chains together with its circuit implementation. The contact resistances of via connections are also shown corresponding to the layout design. (b) Narrow armchair-edge GNRs and wide zigzag-edge GNRs are used as channel material and local interconnects, respectively. (c) 3D view of a GNRFET with one ribbon of armchair GNR(*N*,0) as channel material.

GNR is atomically thin in vertical direction and quantum-mechanically confined in transverse direction, leading to a small drive current of GNRFETs. Fabricating multiple parallel armchair

GNRs in an array connected to the same wide zigzag GNRs can increase the drive current, leading to better switching attribute for high-performance applications [22]. As such, the number of GNR channels (N_r) in a GNRFET can correspond to integer increment of W/L in conventional CMOS. As the bandgap of GNRs can be inversely changed by the GNR width (W_{GNR}), it provides another degree of freedom for a designer to use bandgap engineering in GNRFET circuits [23]. The length of local GNR interconnects between transistors within logic gates are much shorter than the mean free path of graphene. In addition, the GNR interconnects are wide enough to maintain the condition of very small bandgap. Therefore, the effects of edge roughness on interconnects can be neglected. Also the resistance and capacitance of local GNR interconnects can be assumed negligible in first-order models [24]. The multi-channel GNRFET can be interpreted as several parallel individual GNRFETs, as shown in Fig. 2.3c. We will develop both computational and analytical models for this GNRFET in the next chapters.

References

1. Moore, G. E. (2006). Cramming more components onto integrated circuits, Reprinted from Electronics, volume 38, number 8, April 19, 1965, p. 114, *IEEE Solid-State Circuits Newsletter*, **3**, pp. 33–35.
2. Plummer, J. D. and Griffin, P. B. (2001). Material and process limits in silicon VLSI technology, *Proceedings of the IEEE*, **89**, pp. 240–258.
3. International Technology Roadmap for Semiconductors (ITRS) (2013), http://www.itrs.net/.
4. Schulz, M. (1999). The end of the road for silicon? *Nature*, **399**, pp. 729–730.
5. Srivastava, A. and Banadaki, Y. M. (2016). Graphene transistors: Present and beyond, in: *Nano-CMOS and Post-CMOS Electronics: Devices and Modelling*, S. P. Mohanty and A. Srivastava (Eds.), The Institution of Engineering and Technology, pp. 99–137.
6. Iyechika, Y. (2010). Application of graphene to high-speed transistors: Expectations and challenges, *Science and Technology Trends*, **37**, pp. 76–92.
7. Schwierz, F. (2013). Graphene transistors: Status, prospects, and problems, *Proceedings of the IEEE*, **101**, pp. 1567–1584.

8. Harada, N., Sato, S., and Yokoyama, N. (2013). Theoretical investigation of graphene nanoribbon field-effect transistors designed for digital applications, *Japanese Journal of Applied Physics*, **52**, p. 094301.

9. Avouris, P., Chen, Z., and Perebeinos, V. (2007). Carbon-based electronics, *Nature Nanotechnology*, **2**, pp. 605–615.

10. Johari, Z., Hamid, F., Tan, M. L. P., Ahmadi, M. T., Harun, F., and Ismail, R. (2013). Graphene nanoribbon field effect transistor logic gates performance projection, *Journal of Computational and Theoretical Nanoscience*, **10**, pp. 1164–1170.

11. Kim, K., Choi, J.-Y., Kim, T., Cho, S.-H., and Chung, H.-J. (2011). A role for graphene in silicon-based semiconductor devices, *Nature*, **479**, pp. 338–344.

12. Novoselov, K. S., Geim, A. K., Morozov, S. V., Jiang, D., Zhang, Y., Dubonos, S. V., Grigorieva, I. V., and Firsov, A. A. (2004). Electric field effect in atomically thin carbon films, *Science*, **306**, pp. 666–669.

13. Banadaki, Y. M., and Srivastava, A. (2015). Scaling effects on static metrics and switching attributes of graphene nanoribbon FET for emerging technology, *IEEE Transactions on Emerging Topics in Computing*, **3**, pp. 458–469.

14. Obradovic, B., Kotlyar, R., Heinz, F., Matagne, P., Rakshit, T., Giles, M., Stettler, M., and Nikonov, D. (2006). Analysis of graphene nanoribbons as a channel material for field-effect transistors, *Applied Physics Letters*, **88**, pp. 142102.

15. Rakheja, S., Kumar, V., and Naeemi, A. (2013). Evaluation of the potential performance of graphene nanoribbons as on-chip interconnects, *Proceedings of the IEEE*, **101**, pp. 1740–1765.

16. Todri-Sanial, A., Dijon, J., and Maffucci, A. (Eds.) (2017) *Carbon Nanotubes for Interconnects: Process, Design and Applications*, Springer International Publishing.

17. Berger, C., Song, Z., Li, X., Wu, X., Brown, N., Naud, C., Mayou, D., Li, T., Hass, J., and Marchenkov, A. N. (2006). Electronic confinement and coherence in patterned epitaxial graphene, *Science*, **312**, pp. 1191–1196.

18. Van Noorden, R. (2006). Moving towards a graphene world, *Nature*, **442**, pp. 228–229.

19. Banadaki, Y. M., and Srivastava, A. (2015). Investigation of the width-dependent static characteristics of graphene nanoribbon field effect transistors using non-parabolic quantum-based model, *Solid-State Electronics*, **111**, pp. 80–90.

20. Kang, J., Sarkar, D., Khatami, Y., and Banerjee, K. (2013). Proposal for all-graphene monolithic logic circuits, *Applied Physics Letters*, **103**, p. 083113.
21. Nagashio, K., Nishimura, T., Kita, K., and Toriumi, A. (2009). Metal/graphene contact as a performance killer of ultra-high mobility graphene analysis of intrinsic mobility and contact resistance, *2009 IEEE International Electron Devices Meeting (IEDM)*, Baltimore, MD, pp. 1–4.
22. Chen, Z., Lin, Y.-M., Rooks, M. J., and Avouris, P. (2007). Graphene nano-ribbon electronics, *Physica E: Low-Dimensional Systems and Nanostructures*, **40**, pp. 228–232.
23. Han, M. Y., Özyilmaz, B., Zhang, Y., and Kim, P. (2007). Energy band-gap engineering of graphene nanoribbons, *Physical Review Letters*, **98**, p. 206805.
24. Choudhury, M. R., Yoon, Y., Guo, J., and Mohanram, K. (2011). Graphene nanoribbon FETs: Technology exploration for performance and reliability, *IEEE Transactions on Nanotechnology*, **10**, pp. 727–736.

Chapter 3

Computational Carrier Transport Model of GNRFET

As the fabrication technology of graphene nanoribbon field-effect transistors (GNRFETs) is still at an early stage, the significant progress in experiments and the evaluation of device performance are accompanied with substantial achievements in theoretical work based on computational simulation techniques. The GNRFET has been simulated by solving a quantum transport model based on non-equilibrium Green's function (NEGF) formalism that fully treats short-channel-length electrostatic effects and quantum tunneling effects. The model will be used to study the static metrics and switching attributes of GNRFET in the next chapters.

3.1 Introduction

Three approaches based on classical, semi-classical, and quantum mechanics can be used for the study of current transport in transistors. The classical approaches are based on Newton's law [1, 2], like charge-collection equations [3] or drift-diffusion equations, which can be employed to model transistors of large dimensions, but is not suitable for the physical modeling of sub-nanometer channel length of MOSFET (metal-oxide semiconductor FET) types due to the significance of quantum effects. The traditional approach usually

Graphene Nanostructures: Modeling, Simulation, and Applications in Electronics and Photonics
Yaser M. Banadaki and Safura Sharifi

ISBN 978-981-4800-36-5 (Hardcover), 978-0-429-02221-0 (eBook)
www.jennystanford.com

focuses on scattering effects inside the channel because of diffusive motions of carriers, whose length is much longer than the mean free path of carriers, as shown in Fig. 3.1.

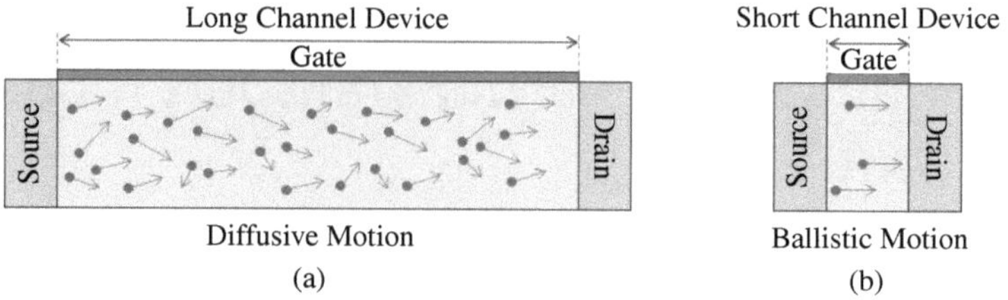

Figure 3.1 (a) Diffusive carrier transport in long-channel devices. (b) Ballistic carrier transport in short-channel devices. Reprinted from Ref. [4], Copyright 2015, with permission from Elsevier.

Figure 3.2a shows the typical structure of MOSFET-like GNRFETs. In this structure, a semiconducting armchair GNR is sandwiched between two thin insulator layers in a double metal gate topology. Here we will focus on studying the carrier transport in nonmetric transistors to develop a quantum transport algorithm. In the next chapters, we will study the effect of actual device dimension and material properties.

Figure 3.2b shows the energy-position-resolved local density of states (LDOS), which is numerically simulated by a quantum-based model [4]. LDOS is a physical quantity that describes the density of states (DOS), but at different points in space, and is then a function of energy and position. Similar results as computational methods can be obtained by scanning tunneling spectroscopy (STS), which is capable of imaging electron densities of states as a function of energy at a given location in the sample. In the figure, the bandgap with quite low LDOS (dark black region) and the channel potential barriers can be easily identified. The quantum interference pattern due to incident and reflected electron waves in the generated quantum well in the valence band of the channel is also apparent. The carrier transport can be associated with three mechanisms: (1) thermionic current for electrons emission above the channel potential barrier, (2) direct source-to-drain tunneling current through channel potential barrier, and (3) band-to-band electron tunneling from the channel to the drain regions.

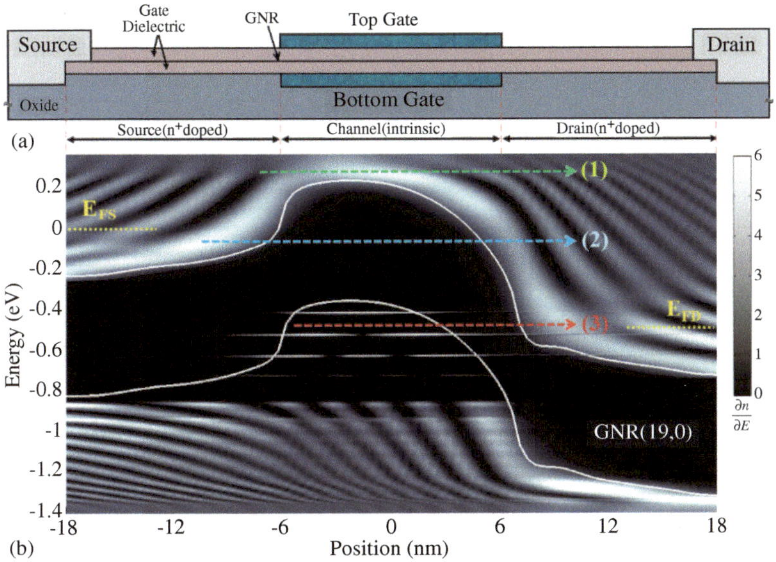

Figure 3.2 (a) Typical structure of MOSFET-like graphene nanoribbon field-effect transistors. (b) Energy-position-resolved local density of states of a typical GNRFET simulated with NEGF formalism, showing three possible regions for carrier transport: (1) thermionic emission of carriers over the channel potential barrier, (2) direct tunneling of carriers through channel potential barrier, and (3) band-to-band tunneling of electrons from valence band in the channel region to the empty states in the drain side [5].

For an emerging device such as GNRFET, the channel length needs to be 10 nm or less and the mean free path can reach to a few micrometers. Thus, the transport can be interpreted as ballistic motion of carriers in short-channel devices, while the discrete energies of GNR channel can be tuned by the gate electrostatic potential, leading to important effects of quantum tunneling on carrier transport. It is shown in Fig. 3.2b that the direct source-to-drain tunneling and band-to-band tunneling from drain to channel can be significant by scaling down the channel length and the width of GNR, respectively. The semi-classical models [6–8] can be modified to incorporate band-to-band tunneling current, but the models cannot be used for GNRFET with channel length below 10 nm because the direct carrier tunneling from source to drain regions can be an important component in calculating the drain-to-source current. Thus, the atomistic quantum-based models [9, 10] need to be used in order to investigate the GNRFET performance by scaling down the channel length because it takes into consideration the tunneling effects in short-channel GNRFET.

The quantum-based simulation is the most computationally demanding approach as the quantum effects become more and more important by scaling down the channel length. The most accurate quantum-based method for bottom-up device simulation is the NEGF approach, where the Schrödinger equation is solved under non-equilibrium condition. NEGF formalism provides the atomistic description of channel material as well as the effects of contacts and scattering on carrier transport in the channel. The discretization of device Hamiltonian provides two alternative approaches for applying NEGF formalism: real space formulation [11], which can be used directly for any geometry, and mode space formulation [12], which splits up the device simulation into a set of one-dimensional (1D) problems over sub-bands. The mode space approach can be applied for the simulation of GNRFET by assuming smooth edges and negligible potential variation in the transverse direction. It has been successfully applied for simulating a variety of nanometer channel materials such as carbon nanotube [13, 14], silicon MOSFET [15], and graphene nanoribbon [16, 17].

There is not much reported work on the scaling of GNRFETs, especially below 10 nm channel length, in which direct tunneling through channel potential barrier can be significant. Yoon et al. [18] investigated the scaling behavior of graphene-based transistors by performing quantum transport simulations, but limited the scaling down to 30 nm channel length. With the same simulation approach, Ouyang et al. [9] performed a comprehensive study on the scaling behavior of GNRFETs down to 10 nm. Similarly, research on the width-dependent study of GNRFET with respect to GNR index is also limited. Ouyang et al. [9] showed the scaling behavior of GNRFETs considering only one semiconducting family of armchair GNRs. Raza and Kau [19] classified armchair GNRs into three families but considered only the bandgap and the effective mass of the first sub-band. Sako et al. [20] investigated the effects of edge bond relaxation in GNRFET with 10 nm channel length by considering only the effective mass of the first sub-band in top-of-the barrier model. Kliros [21] studied the effect of the width-dependent performance of GNRFETs using an analytical model.

The variety of carrier transport models can be used to evaluate the performance of GNRFETs, including either simplified semi-classical transport models [6, 7, 20] or quantum transport models [9, 16]. However, the traditional approach is not suitable as the assumption of significant scattering inside the channel is not valid for short-channel length as the mean free path for carriers is much smaller than the channel length [22]. The mean free path in smooth-edge nanoribbons is around hundreds of nanometers at room temperature due to the weak electron–phonon interaction [23]. As such, performance studies of armchair GNR families with channel length below 10 nm are to be researched and a more comprehensive investigation is thus warranted based on more sophisticated approaches. The other classical and semi-classical methods cannot treat short gate-length electrostatic effects and quantum tunneling effects such as direct source-to-drain tunneling in short-channel GNRFET or band-to-band tunneling at the source and drain junctions [24]. In addition, the existence of mismatch between the parabolic band approximation and the exact dispersion relation in analytical models [21], top-of-the-barrier model [8], or semi-analytical model [25] may not correctly estimate the actual concentration of carriers in the channel.

Scaling down the channel length increases the quantum effects and thereby a bottom-up quantum-based simulation is vitally important approach to gain insight into GNRFET performance in sub-10 nm channel length. NEGF formalism incorporates the atomistic description of channel material and the effects of contacts on carriers transport in the channel. In principle, there are two alternative approaches to use NEGF formalism for nanoscale device simulation: the real space formalism, which is directly applicable to any geometry, and the mode space formalism, which splits up the 2D GNR into a set of 1D problems corresponding to the generated sub-bands due to structural confinement in the transverse direction. The mode space approach can be applied for the GNRFET simulation assuming the perfect edges and thus negligible potential variation in the transverse direction. This assumption results in a considerable computational advantage while maintaining the accuracy of device simulation.

3.2 Quantum Transport Model

A full quantum transport model based on NEGF formalism in mode space is developed for the simulation of GNRFET [26, 27], where the energy-position-dependent Hamiltonian is employed using non-parabolic effective mass (NPEM) model [28]. The existence of mismatch between the parabolic band approximation and the exact dispersion relation in analytical models [21], top-of-the-barrier model [8], or semi-analytical model [25] may not correctly estimate the actual concentration of carriers in the channel. In this section, the quantum transport model of GNRFETs will be developed and used for investigating the scaling of its channel length down to 2.5 nm, as well as the width-dependence performance of GNRFETs with respect to GNR index.

Figure 3.3a shows an iterative procedure between electrostatic and transport solutions, in which calculating the potential profile depends on the carrier density and calculating the carrier density needs the potential profile along the device. As such, before calculating the drain-to-source current for a bias condition, the potential profile and charge density need to be obtained by constructing a self-consistent calculation between Poisson equation and transport equations. Figure 3.3b illustrates a flowchart for the detail of self-consistent algorithm, which has been explained in following seven steps.

Step I: For a given width, the effective masses of the lowest sub-bands have been extracted by tight-binding (TB) calculation for a slab with zero potential for use in the successive transport calculations of the self-consistent loop.

For obtaining GNR dispersion relation, TB calculation can be employed based on the nearest-neighbor orthogonal p_z orbitals as basis functions. One p_z orbital is enough for the atomistic physical description of graphene since energy levels of s, p_x, and p_y orbitals are far from the Fermi level and do not play important roles for carrier transport. Figure 3.4a shows the atomic view of armchair-edge GNR(N_a,0), where ribbon index $N_a = 13$ is the number of dimer lines in the transverse direction. The GNR width is commonly defined as $W_{GNR} = (N_a - 1)\sqrt{3}\,a_{cc}/2$, where a_{cc} is the carbon–carbon bonding length. Calculating TB inside the slabs with the length of $3a_{cc}$ and $2N_a$ atoms can give the required information of GNR sub-bands [29] for the transport calculation. The matrix element of the Hamiltonian between the αth atom within the nth slab and the βth atom within the mth slab is written as follows:

$$H_{n\alpha,m\beta} = H^{0}_{n\alpha,m\beta} + \delta_{n\alpha,m\beta} U_{n\alpha} \tag{3.1}$$

where $\delta_{n\alpha,m\beta}$ is the Kronecker delta and $U_{n\alpha}$ is the electrostatic potential energy at the (n,α) atom site. $H^{0}_{n\alpha,m\beta}$ is equal to the nearest-neighbor hopping energy, $t = -2.7$ eV if the atoms (n,α) and (m,β) are first nearest neighbors and equal to zero otherwise. The graphene lattice has been abruptly terminated at the edge and occupied by hydrogen atoms, which can be modeled if the hopping energy for pairs of atoms along the edges of the GNR is assumed $t(1 + \gamma)$ for the correction factor of $\gamma = 0.12$ [30]. The TB model of edge bond relaxation has been verified by the first principle calculations showing the identical results for the band structure of GNR near the Fermi level [28]. The edge bond relaxation has a significant effect on both the bandgap energy and the effective mass of GNR sub-bands [30].

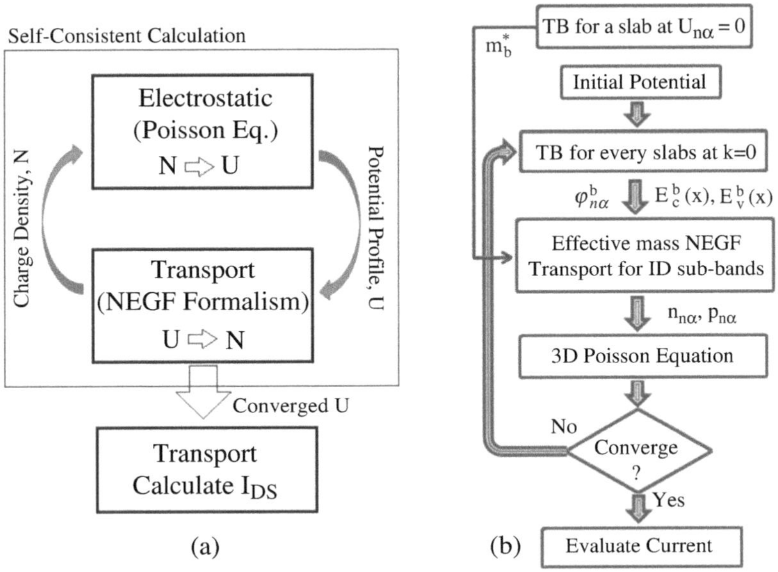

Figure 3.3 (a) Self-consistent calculation between electrostatic (Poisson equation) and transport (NEGF formalism) solutions. (b) Flowchart for the self-consistent algorithm as described in Step I through Step VII in the text. Reprinted from Ref. [4], Copyright 2015, with permission from Elsevier.

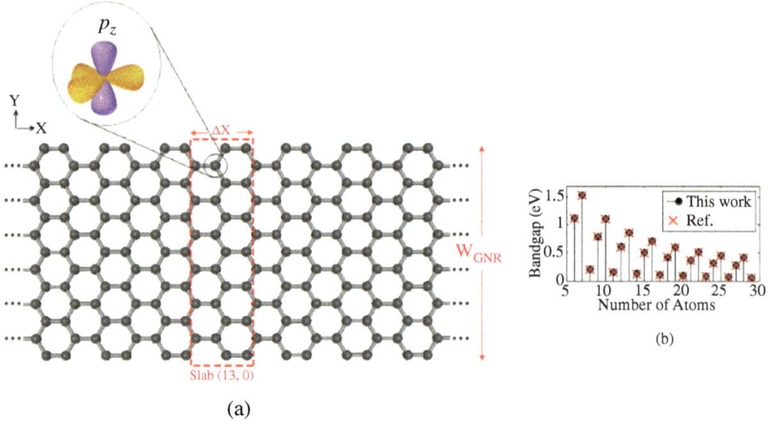

Figure 3.4 (a) Schematic cross section of an armchair GNR(13,0) and the corresponding slab used in TB calculation in the transverse direction. (b) The bandgap energy of three GNR families versus the GNR index. *Note*: The calculated bandgap energy in this work has been compared with those of Ref. [20].

Figure 3.4b shows the close agreement of the calculated bandgap energy in this work with those of Sako et al. [20]. The quantum confinement of carriers in one dimension can open the bandgap at the expense of reducing the electron velocity and degrading the band linearity near the Dirac point. The nonlinearity can be corrected for each sub-band using an effective mass model given by [28]

$$\left(E_b(k)-\frac{E_g^b}{2}\right)\left(\frac{1}{2}+\frac{E_b(k)}{E_g^b}\right)=\frac{\hbar^2k^2}{2m_b^*} \tag{3.2}$$

where E_g^b is the energy gap, $E_b(k)$ is the energy, and m_b^* is the effective mass for a sub-band index b. The TB band diagram, the NPEM model, and the constant effective mass model of GNR(7,0), GNR(25,0), GNR(6,0), and GNR(24,0) are shown in Fig. 3.5. The difference between the two models is increased by increasing the GNR width.

Figures 3.6a and 3.6b show the overview of NEGF simulation and an example of Poisson solution in a transistor, respectively, which are referred later in the next steps.

Step II. Considering an initial potential distribution $U_{n\alpha}^{\text{in}}$, the extrema energies $E_c^b(x)$ and $E_v^b(x)$ as well as wavefunction $\varphi_{n\alpha}^b(x)$ for the sub-band index b are obtained as a function of longitudinal direction by repeating TB calculation for every slab of the ribbon only at $k = 0$.

Step III. The Hamiltonian matrix $H_b(E)$, Green's function $G_b(E)$, contacts self-energies $\Sigma^b_{\mathrm{S/D}}(E)$, and the corresponding level broadening function $\Gamma^b_{\mathrm{S/D}}(E)$ have been obtained for a given sub-band, b, where E is the electron energy.

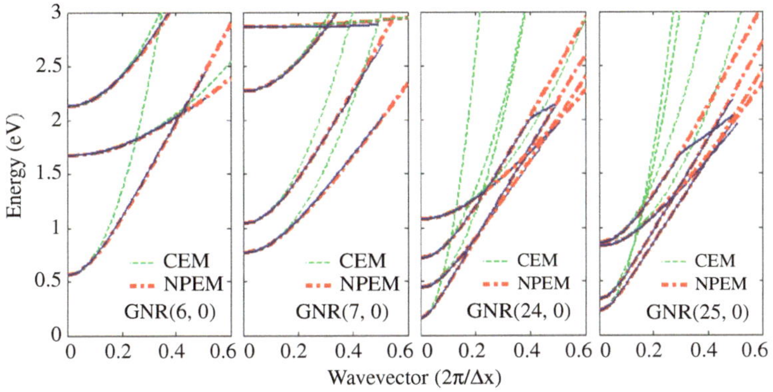

Figure 3.5 Band structure of four members of semiconducting GNRs near the charge neutrality point along with the curves of non-parabolic effective mass model and constant (parabolic) effective mass model. The blue line is the energy obtained from tight-binding calculation. Reprinted from Ref. [4], Copyright 2015, with permission from Elsevier.

The transport equations based on NEGF formalism has a Hamiltonian similar to the TB case with the 1D discretization step equal to the slab width $\Delta X = 3a_{\mathrm{cc}}$, in which the non-parabolic band diagram has been corrected by constructing a position-energy-dependent effective mass model as follows [28]:

$$m_b(x,E) = \begin{cases} m_b^*\left[1 + \dfrac{E - E_c^b(x)}{E_g^b(x)}\right] & \text{if } E > E_i^b(x) \\[2ex] m_b^*\left[1 + \dfrac{E_v^b(x) - E}{E_g^b(x)}\right] & \text{if } E < E_i^b(x) \end{cases} \tag{3.3}$$

where $E_i^b(x)$ is a mid-gap energy and m_b^* is the effective mass for a sub-band index b calculated from Eq. (3.2). Based on the obtained Hamiltonian, the retarded Green's function is constructed as follows:

$$G_b(E) = [EI - H_b - \Sigma_{\mathrm{S}}^b - \Sigma_{\mathrm{D}}^b]^{-1} \tag{3.4}$$

where E is energy, I is identity matrix, Σ_S^b and Σ_D^b are the self-energy matrices of source and drain contacts as shown in Fig. 3.6a, which incorporates the effect of the contacts on channel sub-bands. The self-energy matrices have the same dimension as the Hamiltonian $N \times N$, where N is the number of slabs in the longitudinal direction. For the Hamiltonian with scalar elements, the only non-null elements of the matrices are $\Sigma_S^b(1,1)$ and $\Sigma_D^b(N,N)$, which have been obtained using the piecewise equation in Ref. [31] as follows:

$$\Sigma_S^b(1,1)/t = \begin{cases} (x-1)+\sqrt{x^2-2x} & -\infty \le x \le 0 \\ (x-1)-i\sqrt{2x-x^2} & 0 \le x \le 2 \\ (x-1)-\sqrt{x^2-2x} & 2 \le x \le \infty \end{cases} \tag{3.5}$$

where,

$x = (E - E_c^b(1))/2t$ if $E > E_i^b(1)$ and otherwise $x = (E_V^b(1) - E)/2)$. The parameter t and energy $E_i^b(1)$ are the nearest-neighbor hopping energy and mid-gap energy at the first slab on the source side, respectively. The Hamiltonian H_b depends on energy through the position-dependent effective mass in Eq. (3.3).

Before connecting the GNR channel to the source and drain contacts, the DOS of the GNR channel consists of sharp levels at the sub-band minimum energies due to quantum confinement, while there is a continuous distribution of states in the source and drain contacts. Coupling the discretized states in the channel to the continuous states in the contacts, part of the sharp states in the channel spreads into contacts and part of the contact states spreads into the channel. As such, the initial sharp structures of the DOS of the GNR channel spread out over a range of energies and broaden around the initial sharp levels. The level broadening quantities Γ_S and Γ_D for a sub-band b can be calculated as follows:

$$\Gamma_S^b = i(\Sigma_S^b - \Sigma_S^{b+}) \tag{3.6}$$

$$\Gamma_D^b = i(\Sigma_D^b - \Sigma_D^{b+})$$

where i and the superscript '+' refer to the imaginary unit and the Hermitian transpose operator, respectively.

Step IV: Calculate the source and drain correlation functions $G_b^<(E)$ and $G_b^>(E)$ as well as the corresponding electron number $n_{l\alpha}$ and hole number $p_{l\alpha}$.

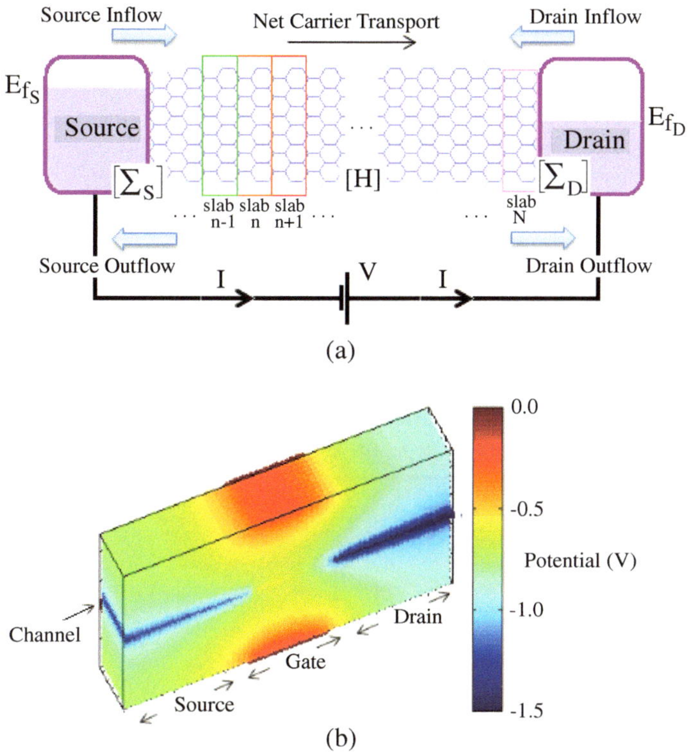

Figure 3.6 (a) Conceptual sketch of the armchair edge GNR channel, including the quantities used in the NEGF formalism. (b) Example of 3D potential distribution calculated by solving 3D Poisson equation [5].

The electron and hole correlation functions can be calculated by

$$G_b^{<}(E) = G_b(E)[\Sigma_S^{b<}(E) + \Sigma_D^{b<}(E)]G_b^{+}(E)$$

$$G_b^{>}(E) = G_b(E)[\Sigma_S^{b>}(E) + \Sigma_D^{b>}(E)]G_b^{+}(E) \quad (3.7)$$

where $\Sigma_{S/D}^{b<}(E)$ is the inflow of carriers from the source and drain contacts into the channel region for sub-band b, as shown in Fig. 3.1b. Similarly, $\Sigma_{S/D}^{b>}(E)$ is the outflow of carriers from the channel region into the source and drain contacts for sub-band b. These quantities depend on the condition in the contacts (Fermi levels) and the channel coupling to the contacts (level broadening), which can be obtained as follows:

$$\Sigma_{\text{S/D}}^{b<}(E) = i\Gamma_{\text{S/D}}^{b} f_{\text{S/D}}(E) \tag{3.8}$$

$$\Sigma_{\text{S/D}}^{b>}(E) = i\Gamma_{\text{S/D}}^{b}[1 - f_{\text{S/D}}(E)] \tag{3.9}$$

where $f_{\text{S/D}}$ (E) is the Fermi functions of source and drain contacts as follows:

$$f_{\text{S/D}}(E) = \left[1 + \exp\left(\frac{E - E_{\text{F}_{\text{S/D}}}}{k_{\text{B}}T}\right)\right]^{-1} \tag{3.10}$$

In Eq. (3.10), k_{B} is the Boltzmann constant, $E_{\text{F}_\text{S}} = E_{\text{F}}$ and $E_{\text{F}_\text{D}} = E_{\text{F}} - qV_{\text{DS}}$ are the Fermi levels of the source and drain contacts, respectively, as shown in Fig. 3.6a. E_{F} is the reference Fermi level of GNR, and V_{DS} is the applied drain-to-source voltage. The electron and hole numbers at the (n,α) atom site, where α is the index of atom in the nth slab, can be achieved by summations over all sub-bands as follows:

$$n_{n\alpha} = -2i\sum_{b}\left[\left|\varphi_{n\alpha}^{b}\right|^{2} \int_{E_i^b(x)}^{\infty} \frac{1}{2\pi} G_b^{<}(n,n;E)dE\right] \tag{3.11}$$

$$p_{n\alpha} = 2i\sum_{b}\left[\left|\varphi_{n\alpha}^{b}\right|^{2} \int_{-\infty}^{E_i^b(x)} \frac{1}{2\pi} G_b^{>}(n,n;E)dE\right] \tag{3.12}$$

Step V: Insert the electron/hole numbers into the Poisson equation to obtain a new potential energy $U_{n\alpha}$.

The actual potential inside the channel in response to the voltages applied to the external electrodes is required to calculate the full current–voltage characteristics. In order to obtain the electrostatic potential energy $U_{n\alpha}(\vec{r})$ and use it as the diagonal entry of the TB Hamiltonian matrices in Step II, the 3D Poisson equation is solved as follows:

$$\nabla.[\varepsilon(\vec{r})\nabla U_{n\alpha}(\vec{r})] = qQ(\vec{r}) \tag{3.13}$$

where $\varepsilon(\vec{r})$ is the permittivity of dielectric materials, q is electron charge, and $Q(\vec{r})$ is the net charge density distribution determined by the doping profile and the calculated electron and hole numbers of GNR channel. Considering the profile of charge density and the potentials at electrodes, the Poisson equation is solved using the finite difference method by considering Drichlet boundary condition [32] at the metallic gate electrodes $U = V_G$ and Neumann boundary condition [32] at the remaining boundaries, e.g., dielectric materials, where the electric field perpendicular to the boundary is assumed to be zero. Figure 3.6b shows an example of Poisson solution in a typical MOSFET-like GNRFET.

Step VI: Check the convergence condition: $\left|U_{n\alpha} - U_{n\alpha}^{\text{old}}\right| < \xi$. If yes, go to the next step; otherwise replace $U_{n\alpha}^{\text{old}}$ by the calculated $U_{n\alpha}$ and go to Step II.

Finally, the total current can be calculated as follows:

$$I_{\text{DS}} = \frac{q^2}{h} \int_{-\infty}^{\infty} \sum_b \frac{4}{q} \Re\left\{H_b(n,n+1;E) G_b^{<}(n+1,n;E)\right\} dE \tag{3.14}$$

where h is the Planck constant and symbol $\Re$ indicates the real part. Considering coherent transport, the equation can be reduced to Landauer formalism [31] as follows:

$$I_{\text{DS}} = \frac{2q}{h} \int_{-\infty}^{\infty} T(E)[f_{\text{S}}(E) - f_{\text{D}}(E)] dE \tag{3.15}$$

where $T(E) = \sum_b T_b(E)$ is the total transmission coefficient with $T_b(E)$ being the transmission coefficient of the bth sub-band described by

$$T_b(E) = \text{Trace}[\Gamma_{\text{S}}^b G_b \Gamma_{\text{D}}^b G_b^{+}] \tag{3.16}$$

where G^{+} is the advanced Green's function.

3.3 Quantum Capacitance in GNRFET

The total charge density Q can be obtained by summing the electron and hole densities in the channel from Eq. (3.11) and Eq. (3.12). Figure 3.7a shows an example of charge density and the corresponding 1D potential profile as a function of position along the longitudinal direction, which has been simulated using NEGF formalism. As edge states are small in armchair GNRs and consequently charge distribution in the transverse direction is uniform, the electrostatic potential on the GNR and voltage drop over the gate oxide are also uniform [33]. Thus, the gate voltage V_G is simply the summation of the voltage drop over the gate oxide V_{ox} and the electrostatic potential on the GNR, V_S, leading to the expression in Eq. (3.17).

$$\frac{dV_G}{dQ} = \frac{dV_S}{dQ} + \frac{dV_{ox}}{dQ} \tag{3.17}$$

By defining, the quantum capacitance as $CQ = dQ/dV_S$ and the gate insulator capacitance per unit area as $C_{ins} = dQ/dV_{ox}$, the total gate-to-source capacitance C_G can be obtained as follows:

$$\frac{1}{C_G} = \frac{1}{C_Q} + \frac{1}{C_{ins}} \tag{3.18}$$

where C_{ins} is given by

$$C_{ins} = N_G \kappa \varepsilon_0 \left(\frac{W_G}{t_{ins}} + \alpha \right) \tag{3.19}$$

where N_G is the number of gates, equal to 2 for the DG geometry; κ is the relative dielectric constant of the insulator material; t_{ins} is the gate insulator thickness; W_G is the width of the gate metal contact set equal to the GNR width W_{GNR} in the simulation; and $\alpha = 1$ is a dimensionless fitting parameter due to the electrostatic edge effect. The gate insulator capacitance increases linearly with GNR width due to the increase in the area of GNR [33]. The effective gate-to-source capacitance is obtained by the series combination of insulator capacitance and quantum capacitance, as shown in Fig. 3.7b. In order to have the same gate electrostatic control on the channel by scaling down the gate length, the strategy was to scale down the insulator thickness for decades [34]. In typical silicon MOSFETs, the gate insulator capacitance is smaller, and thereby it is the dominant factor in calculating the equivalent gate-to-source capacitance.

For nanostructures such as GNR, the carriers exhibit 1D transport, and the corresponding DOS is very low (see Fig. 3.4c) because it is atomically thin in the vertical direction and quantum mechanically confined in the transverse direction. Thus, the quantum capacitance of GNRFET can be very small, such that the total gate-to-source capacitance of a GNRFET is dominantly determined by the quantum capacitance of GNR. Hence, increasing insulator capacitance cannot make a significant increase in equivalent gate-to-source capacitance of GNRFET at quantum capacitance limit (QCL). The application of high-k gate dielectrics, high-geometry gate, and scaling down the insulator thickness can significantly increase the insulator (or geometrical) capacitance C_{ins} and thereby strongly promote the device operation close to QCL [35]. The assumption of QCL and neglecting C_{ins} is exclusively correct for long-channel GNRFET as the channel potential energy is dominantly controlled by the gate electrode and a simple analytical closed-form model can be developed [36]. By scaling the channel, however, the drain and source voltages can change the potential profile and the corresponding charges in the channel, especially when the quantum capacitance is increased at on-state. Thus, the full dominance of quantum capacitance may not be an accurate assumption. In addition, the DOS of GNR and the corresponding quantum capacitance as a function of gate voltage can be also altered by scaling the width of GNR in GNRFET [21]. Thus, the numerical simulation is required for the accurate investigation of GNRFET performance [9].

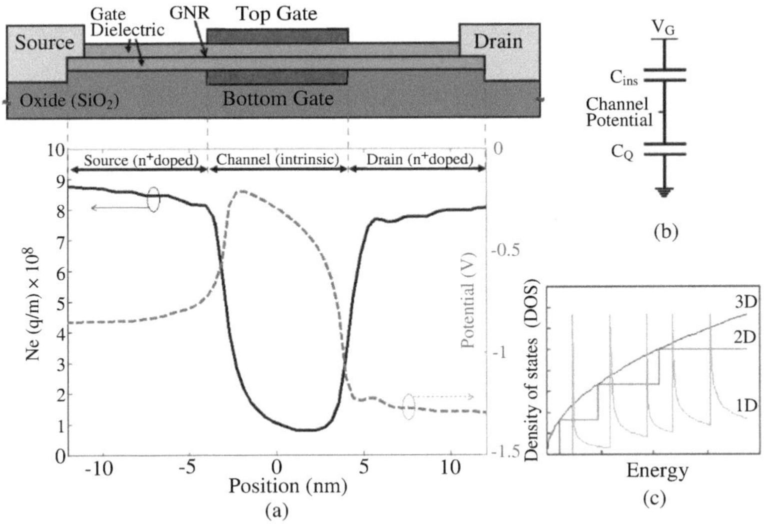

Figure 3.7 (a) An example of 1D potential profile (right axis) and the charge density per unit length as a function of position along the longitudinal direction, computed for GNR(13,0) at V_{DS} = 0.5 V and V_{GS} = 0.2 V. (b) Series configuration of electrostatic capacitance and quantum capacitance. (c) Density of states versus energy for 1D, 2D, and 3D semiconductors [5].

In the QCL regime, the DOS of a graphene nanoribbon is an important factor, which can alter the channel charge and the corresponding quantum capacitance depending on the relative location of the Fermi level and the position of GNR sub-bands in energy. For instance, the channel charge and the quantum capacitance of GNR(6,0) and GNR(10,0) are shown in Fig. 3.8a and Fig. 3.8b, respectively. Both the GNRs have the same bandgap energy of E_g = 1.1 eV, but they exhibit different quantum capacitance due to the location of upper sub-bands and the difference in their DOS, as shown in Fig. 3.8c and Fig. 3.8d. GNR(6,0) has larger quantum capacitance with steeper increase with increasing gate voltage because the second sub-band of GNR(10,0) is close to the first sub-band and both sub-bands have larger effective masses than the first sub-band of GNR(6,0).

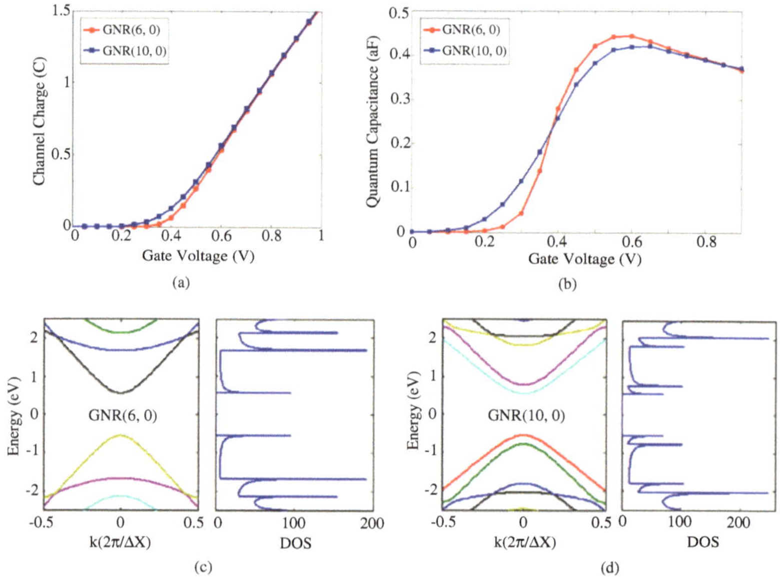

Figure 3.8 (a) Channel charge and corresponding (b) quantum capacitance versus gate voltage for GNR(6,0) and GNR(10,0). *Note*: L_{CH} = 10 nm, V_{DS} = 0.5 V, and the dielectric layer is assumed aluminum nitride (AlN) with the relative dielectric permittivity k = 9. (c) and (d) show the energy dispersion and density of states of two members of GNR families (3p,0) and (3p+1,0), respectively, with approximately same bandgap close to E_g = 1.1 eV [37].

3.4 Computational Time

Accurate results and deeper physical insight can be achieved by atomistic quantum transport models at the expense of long computational time. Yet, a considerable computational advantage and relatively accurate results can be achieved by solving the self-consistent NEGF formalism in mode space basis as has been already employed for the simulation of conventional MOSFETs [32, 37], carbon nanotube FETs [38, 39], and GNRFETs [12, 26]. The transverse confinement of GNR converts the transport problem into a few 1D sub-bands, allowing us to obtain further computational advantage by incorporating only a few lowest sub-bands, which participate in carrier transport within the energy interval under investigation. As can be seen from the charge density and the drain current of four GNRs in Fig. 3.9a and Fig. 3.9b, the third and fourth sub-bands contribute mostly in charge density calculation of wider GNRs and need to be considered in the self-consistent loop. While these have minor effects on the amount of the drain current and charge density of the narrower GNRs, which can be neglected at a range of bias voltage in the width study, leading to a large computational advantage.

The TB computational time is increased by increasing GNR width as shown in Fig. 3.9c. The TB calculation can be very computationally intensive [40] as it needs to be repeated for every slab of the ribbon to extract the sub-band energies and the square moduli of the eigenfunction as a function of longitudinal direction, leading to TB calculation equal to $L_G/3a_{cc}$ times in a self-consistent loop. Thus, for only one bias condition, the required time for TB routine is equal to $TBCT \times (L_G/3a_{cc}) \times N_{SC}$, where N_{SC} is the number of self-consistent loop repetitions and *TBCT* is the TB computational time. Using the NPEM model, the effective masses of the lowest sub-bands have been extracted by only one TB calculation for a slab with zero potential in order to use in the successive self-consistent calculations. Then, in self-consistent field (SCF) loop, the transverse wave functions and the energy profile of sub-bands as a function of longitudinal direction have been obtained for every slab of the ribbon only at wave vector $k = 0$, leading to the computational time equal to $TBCT + TBCT \times (L_G/3a_{cc}) \times N_{SC}/N_E$, where N_E is the number of energy discretization. By increasing the GNR index, this can dramatically increase the computational time with respect to TB model as more sub-bands are required to be considered in transport calculation by decreasing the bandgap. Consequently, the computational time can be very intensive depending on the energy discretization and the bias conditions. The NPEM model can lead to roughly two orders of magnitude saving in computational time for GNRFET simulation with 0.001 eV energy grid.

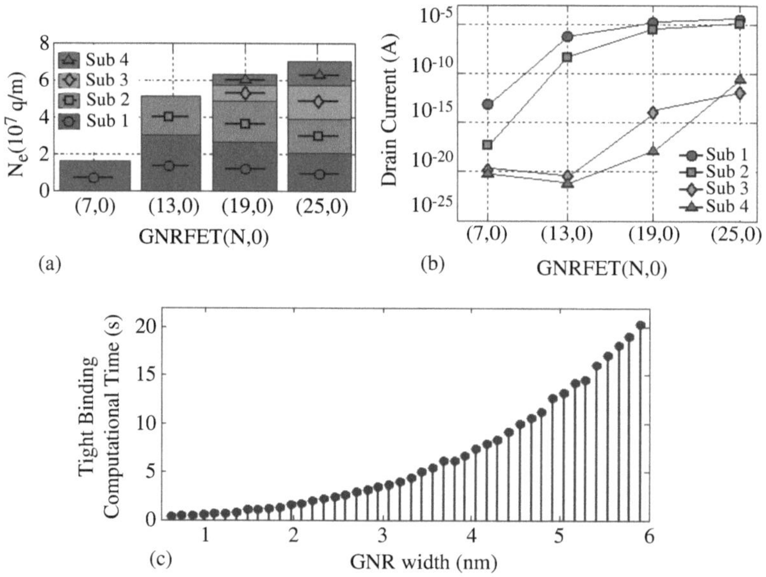

Figure 3.9 Contribution of sub-bands in (a) charge density in the channel, (b) drain current for GNRFET with four GNRs of (7,0), (13,0), (19,0), and (25,0), and (c) tight-binding computational time versus GNR width for energy grid equal to 0.001 eV. Reprinted from Ref. [4], Copyright 2015, with permission from Elsevier.

3.5 Summary

This chapter describes the physical model of carrier transport based on the NEGF formalism. A full quantum transport model based on the NEGF formalism in mode space has been developed for the simulation of GNRFET, where the energy-position-dependent Hamiltonian is employed using the NPEM model. The total gate-to-source capacitance of a GNRFET is dominantly determined by the quantum capacitance of GNR because the quantum capacitance of GNRFET is small. The TB computational time is increased by increasing the GNR width, and thus the NPEM model is developed leading to two orders of magnitude saving in computational time.

References

1. Chen, M.-T. and Wu, Y.-R. (2011). Numerical study of scaling issues in graphene nanoribbon transistors, *MRS Proceedings*, **1344**, doi:10.1557/opl.2011.1352.
2. Ancona, M. G. (2010). Electron transport in graphene from a diffusion-drift perspective, *IEEE Transactions on Electron Devices*, **57**, pp. 681–689.
3. Meric, I., Han, M. Y., Young, A. F., Ozyilmaz, B., Kim, P., and Shepard, K. L. (2008). Current saturation in zero-bandgap, top-gated graphene field-effect transistors, *Nature Nanotechnology*, **3**, pp. 654–659.
4. Banadaki, Y. M. and Srivastava, A. (2015). Investigation of the width-dependent static characteristics of graphene nanoribbon field effect transistors using non-parabolic quantum-based model, *Solid-State Electronics*, **111**, pp. 80–90.
5. Mohammadi Banadaki, Y. (2016). Physical modeling of graphene nanoribbon field effect transistor using non-equilibrium green function approach for integrated circuit design, *LSU Doctoral Dissertations*, 1052.
6. Liang, G., Neophytou, N., Nikonov, D. E., and Lundstrom, M. S. (2007). Performance projections for ballistic graphene nanoribbon field-effect transistors, *IEEE Transactions on Electron Devices*, **54**, pp. 677–682.
7. Ouyang, Y., Yoon, Y., Fodor, J. K., and Guo, J. (2006). Comparison of performance limits for carbon nanoribbon and carbon nanotube transistors, *Applied Physics Letters*, **89**, pp. 203107.
8. Rahman, A., Guo, J., Datta, S., and Lundstrom, M. S. (2003). Theory of ballistic nanotransistors, *IEEE Transactions on Electron Devices*, **50**, pp. 1853–1864.
9. Ouyang, Y., Yoon, Y., and Guo, J. (2007). Scaling behaviors of graphene nanoribbon FETs: A three-dimensional quantum simulation study, *IEEE Transactions on Electron Devices*, **54**, pp. 2223–2231.
10. Guan, X., Zhang, M., Liu, Q., and Yu, Z. (2007). Simulation investigation of double-gate CNR-MOSFETs with a fully self-consistent NEGF and TB method, *IEDM Technical Digest*, **761**, p. 764.

11. Liang, G., Neophytou, N., Lundstrom, M. S., and Nikonov, D. E. (2007). Ballistic graphene nanoribbon metal-oxide-semiconductor field-effect transistors: A full real-space quantum transport simulation, *Journal of Applied Physics*, **102**, p. 054307.
12. Grassi, R., Gnudi, A., Gnani, E., Reggiani, S., and Baccarani, G. (2011). Mode space approach for tight binding transport simulation in graphene nanoribbon FETs, *IEEE Transactions on Nanotechnology*, **10**, pp. 371–378.
13. Javey, A., Guo, J., Wang, Q., Lundstrom, M., and Dai, H. (2003). Ballistic carbon nanotube field-effect transistors, *Nature*, **424**, pp. 654–657.
14. Banadaki, Y. M., Sharifi, S., and Srivastava, A. (2015). Investigation of chirality dependence of carbon nanotube-based ring oscillator, *2015 IEEE 58th International Midwest Symposium on Circuits and Systems (MWSCAS)*, pp. 924–927.
15. Martinez, A., Bescond, M., Barker, J. R., Svizhenko, A., Anantram, M., Millar, C., and Asenov, A. (2007). A self-consistent full 3-D real-space NEGF simulator for studying nonperturbative effects in nano-MOSFETs, *IEEE Transactions on Electron Devices*, **54**, pp. 2213–2222.
16. Fiori, G. and Iannaccone, G. (2007). Simulation of graphene nanoribbon field-effect transistors, *IEEE Electron Device Letters*, **28**, pp. 760–762.
17. Banadaki, Y. M. and Srivastava, A. (2015). Investigation of the width-dependent static characteristics of graphene nanoribbon field effect transistors using non-parabolic quantum-based model, *Solid-State Electronics*, **54**, pp. 462–467.
18. Yoon, Y., Nikonov, D. E., and Salahuddin, S. (2011). Scaling study of graphene transistors, *2011 11th IEEE International Conference on Nanotechnology*, 15–18 August 2011, Portland, OR, USA.
19. Raza, H. and Kan, E. C. (2008). Armchair graphene nanoribbons: Electronic structure and electric-field modulation, *Physical Review B*, **77**, p. 245434.
20. Sako, R., Hosokawa, H., and Tsuchiya, H. (2011). Computational study of edge configuration and quantum confinement effects on graphene nanoribbon transport, *IEEE Electron Device Letters*, **32**, pp. 6–8.
21. Kliros, G. S. (2013). Gate capacitance modeling and width-dependent performance of graphene nanoribbon transistors, *Microelectronic Engineering*, **112**, pp. 220–226.
22. Anantram, M., Lundstrom, M. S., and Nikonov, D. E. (2008). Modeling of nanoscale devices, *Proceedings of the IEEE*, **96**, pp. 1511–1550.

23. Novoselov, K. S., Fal, V., Colombo, L., Gellert, P., Schwab, M., and Kim, K. (2012). A roadmap for graphene, *Nature*, **490**, pp. 192–200.

24. Fiori, G. and Iannaccone, G. (2013). Multiscale modeling for graphene-based nanoscale transistors, *Proceedings of the IEEE*, **101**, pp. 1653–1669.

25. Zhao, P., Choudhury, M., Mohanram, K., and Guo, J. (2008). Analytical theory of graphene nanoribbon transistors, in: *2008 IEEE International Workshop on Design and Test of Nano Devices, Circuits and Systems*, Cambridge, MA, pp. 3–6.

26. Mohammadi Banadaki, Y. and Srivastava, A. (2013). A novel graphene nanoribbon field effect transistor for integrated circuit design, in: *2013 IEEE 56th International Midwest Symposium on Circuits and Systems (MWSCAS)*, Columbus, OH, pp. 924–927.

27. Srivastava, A., Banadaki, Y. M., and Fahad, M. S. (2014). Dielectrics for graphene transistors for emerging integrated circuits, *ECS Transactions*, **61**, pp. 351–361.

28. Grassi, R., Poli, S., Gnani, E., Gnudi, A., Reggiani, S., and Baccarani, G. (2009). Tight-binding and effective mass modeling of armchair graphene nanoribbon FETs, *Solid-State Electronics*, **53**, pp. 462–467.

29. Lin, Y.-M., Perebeinos, V., Chen, Z., and Avouris, P. (2008). Electrical observation of subband formation in graphene nanoribbons, *Physical Review B*, **78**, p. 161409.

30. Son, Y.-W., Cohen, M. L., and Louie, S. G. (2006). Energy gaps in graphene nanoribbons, *Physical Review Letters*, **97**, p. 216803.

31. Datta, S. (2005). *Quantum Transport: Atom to Transistor*, Cambridge University Press.

32. Venugopal, R., Ren, Z., Datta, S., Lundstrom, M., and Jovanovic, D. (2002). Simulating quantum transport in nanoscale transistors: Real versus mode-space approaches, *Journal of Applied Physics*, **92**, pp. 3730–3739.

33. Yoon, Y. and Guo, J. (2007). Effect of edge roughness in graphene nanoribbon transistors, *Applied Physics Letters*, **91**, p. 073103.

34. Bianchi, M., Guerriero, E., Fiocco, M., Alberti, R., Polloni, L., Behnam, A., Carrion, E. A., Pop, E., and Sordan, R. (2015). Scaling of graphene integrated circuits, *Nanoscale*, **7**, pp. 8076–8083.

35. Kliros, G. S. (2012). Scaling effects on the gate capacitance of graphene nanoribbon transistors, in: *CAS 2012 (International Semiconductor Conference)*, Sinaia, pp. 83–86.

36. Jiménez, D. (2008). A current–voltage model for Schottky-barrier graphene-based transistors, *Nanotechnology*, **19**, p. 345204.
37. Jiang, H., Shao, S., Cai, W., and Zhang, P. (2008). Boundary treatments in non-equilibrium Green's function (NEGF) methods for quantum transport in nano-MOSFETs, *Journal of Computational Physics*, **227**, pp. 6553–6573.
38. Pourfath, M., Kosina, H., and Selberherr, S. (2008). Numerical study of quantum transport in carbon nanotube transistors, *Mathematics and Computers in Simulation*, **79**, pp. 1051–1059.
39. Banadaki, Y., Mohsin, K., and Srivastava, A. (2014). A graphene field effect transistor for high temperature sensing applications, *Proc. SPIE 9060, Nanosensors, Biosensors, and Info-Tech Sensors and Systems 2014*, **90600F**, doi: 10.1117/12.2044611.
40. Chin, S.-K., Lam, K.-T., Seah, D., and Liang, G. (2012). Quantum transport simulations of graphene nanoribbon devices using Dirac equation calibrated with tight-binding π-bond model, *Nanoscale Research Letters*, **7**, pp. 1–7.

Chapter 4

Scaling Effects on Performance of GNRFETs

The International Technology Roadmap for Semiconductors (ITRS 2013) has specified the emerging application of alternative channel materials in order to continue the production of a switching transistor for the two categories of high-performance and low-power digital integrated circuits (ICs). In this chapter, we investigate the static device metrics and switching attributes of graphene nanoribbon field-effect transistors (GNRFETs) for scaling the channel length considering OFF-current, I_{ON}/I_{OFF} ratio, subthreshold swing, drain-induced barrier-lowering, voltage transfer characteristic, intrinsic gate capacitance, intrinsic cut-off frequency, intrinsic gate-delay time, and power-delay product (PDP).

4.1 Introduction

To continue the production of a switching transistor, performance has been improved by shortening the gate length by decreasing the capacitance and supply voltage V_{DD}, together with increasing the ON-current, which is characterized by the intrinsic speed of transistor as a guiding metric for projecting the roadmap of emerging technology [1]. There is not much reported work on the scaling of GNRFETs below 10 nm channel length. Yoon et al. [2] investigated

Graphene Nanostructures: Modeling, Simulation, and Applications in Electronics and Photonics
Yaser M. Banadaki and Safura Sharifi

ISBN 978-981-4800-36-5 (Hardcover), 978-0-429-02221-0 (eBook)
www.jennystanford.com

the scaling behavior of graphene-based transistors by performing self-consistent atomistic quantum transport simulations down to 30 nm channel length. With the same simulation approach, Ouyang et al. [3] performed a comprehensive study on the scaling behaviors of GNRFETs down to 10 nm. This chapter investigates the performance and limitation of GNRFETs by reducing the channel length down to 2.5 nm when the vertical scaling of oxide thickness becomes less important by approaching the quantum capacitance limit. The GNRFET structure has been simulated by the self-consistent solution to the three-dimensional (3D) Poisson equation and 1D Schrödinger equation within the non-equilibrium Green's function (NEGF) formalism in mode space, as discussed in Chapter 3. The model can fully treat short-channel-length electrostatic effects and contact effects on the carrier transport in the GNR channel along with the quantum tunneling effects such as direct source-to-drain tunneling in short-channel GNRFET and band-to-band tunneling (BTBT) at the channel–drain junctions in small-bandgap GNRs [4–6].

4.2 Device Structure

Figure 4.1 shows the cross-sectional view and 3D schematic of a double-gate GNRFET structure used in investigating scaling effects. In this structure, the armchair GNR is sandwiched between two thin aluminum nitride (AlN) insulator layers with the relative dielectric permittivity $\kappa = 9$ and the oxide thickness $t_{ins} = 1$ nm in a double metal gate topology. The large-scale and cost-efficient production of thin AlN dielectric layers with good reproducibility and uniformity [7, 8] can result in small equivalent oxide thickness (EOT) while reducing phonon scattering in epitaxial graphene, enabling near ballistic carrier transport in short-channel GNRFETs [9]. The double-gate geometry with high-k dielectric constant offers large gate electrostatic control and consequently large insulator capacitance, which lead to the operation of GNRFETs close to the quantum capacitance limit (QCL) (e.g., $C_{ins} > 10\ C_Q$). While two metals for the source and drain contacts can be directly connected to both sides of an intrinsic GNR channel in a Schottky barrier GNRFET (SB-GNRFET), in MOSFET-type GNRFETs, the extensions of GNR on both sides of the intrinsic channel are needed to be doped in order to tune the carrier injection from the source (drain) reservoirs to the GNR channel [3]. In GNRFETs, drain and source contacts are assumed to be ohmic similar to contacts in conventional MOSFETs. The current is modulated by varying the height of the channel barrier due to the electrostatic potential induced by the applied voltage at the gate. This structure is expected to demonstrate a high I_{ON}/I_{OFF} ratio, outperforming the SB-GNRFET for logic applications [10]. The extensions of source and drain regions with the lengths of L_S and L_D are heavily doped with the concentration of 0.01 n-type dopants per carbon atom and are kept equal to the length of the intrinsic GNR channel in our simulation. The channel between two metallic gates is an intrinsic GNR whose length and width are same as those of the top and bottom gates in the simulation.

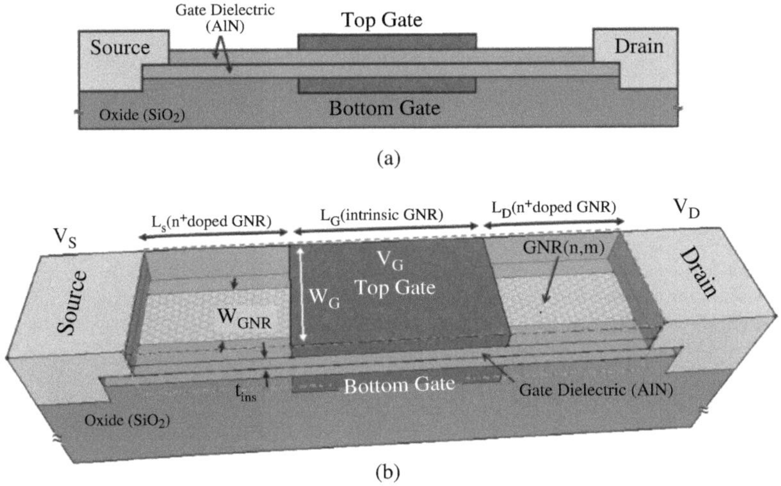

Figure 4.1 (a) Vertical cross section of a double-gate GNRFET. (b) 3D schematic of a double-gate GNRFET structure. Reprinted from Ref. [6], Copyright 2015, with permission from Elsevier.

4.3 Transfer Characteristics of GNRFETs

Figure 4.2a shows the transfer characteristics I_{DS}–V_{GS} for different drain voltages of the GNRFET geometry in Fig. 4.1. For a given drain voltage, the minimum current occurs at the charge neutrality point (CNP), where the hole concentration is equal to the electron concentration and the charge carriers are changed due to the induced electrostatic potential of the gate voltage on sub-bands in the channel. Since the hole mobility is equal to the electron mobility as a result of the symmetric conduction and valence sub-bands, the contribution of electrons and holes in minimum currents is also same, $I_n = I_p = I_{min}/2$. The ambipolar transport is partially recovered with regard to the GNR bandgap as demonstrated experimentally for GNRs with reduced impurity similar to large-area graphene [11].

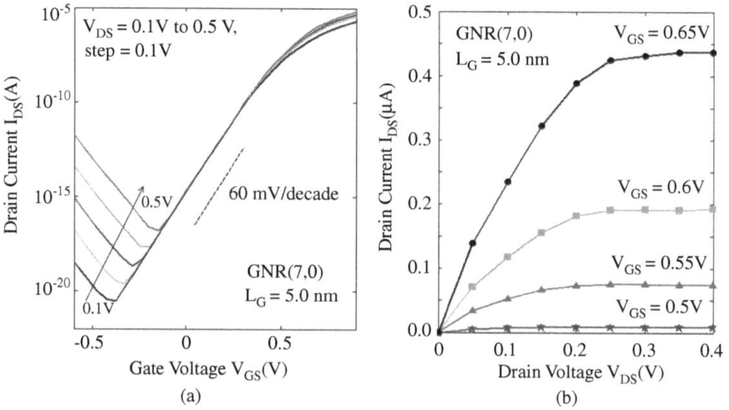

Figure 4.2 (a) I_{DS}–V_{GS} and (b) I_{DS}–V_{DS} of GNR(7,0) with L_G = 5 nm. The test device parameters are given in Fig. 4.1. Reprinted from Ref. [6], Copyright 2015, with permission from Elsevier.

The minimum current is increased and shifted by increasing the drain voltage because the accumulation of holes in the channel is increased because of the BTBT from the source contact to channel together with the drain-induced barrier-lowering (DIBL) effect in short-channel devices. The I_{DS} versus V_{DS} graph for different V_{GS} values of the armchair GNR(7,0) is shown in Fig. 4.2b, which shows strong saturation region in even the short channel length L_G = 5 nm, indicating good MOSFET-type device behavior. It is expected by the ITRS that the saturation drive current of n-MOSFET with channel lengths below 10 nm drops because of the V_{DD} scaling and significant source–drain tunneling [1]. In GNRFETs, the saturation slope mainly depends on the GNR width because increasing V_{DS} in a wider GNR can increase the depletion of electrons in the valence band and, therefore, the accumulation of positive charges in the GNR channel, which can lead to the non-dependence of the saturation region to decreasing channel length [1].

Figures 4.3a,b show the transfer characteristics of GNRFETs for different GNR channel lengths of GNR(7,0) and GNR(13,0), respectively. The bandgaps of GNR(7,0) and GNR(13,0) have been calculated through the tight-binding (TB) method, which results in $E_g = 1.53$ eV and $E_g = 0.86$ eV, respectively. The transconductance curves of GNRFETs are shown in Fig. 4.3b, which indicate the linear dependence on gate voltage after threshold voltage. Downscaling the channel length decreases the gate control on the GNR channel due to short-channel effects, but a more significant factor is GNR width as it can change the size of bandgap, the effective mass of carriers, and the number of available conducting sub-bands in an energy range. This drastically alters BTBT at off-state and the equivalent gate-to-source capacitance of a short-channel device at the QCL. Scaling the gate length shifts the gate voltage at the CNP, which is experimentally interpreted [12] as the signatures of short-channel effects in a graphene device.

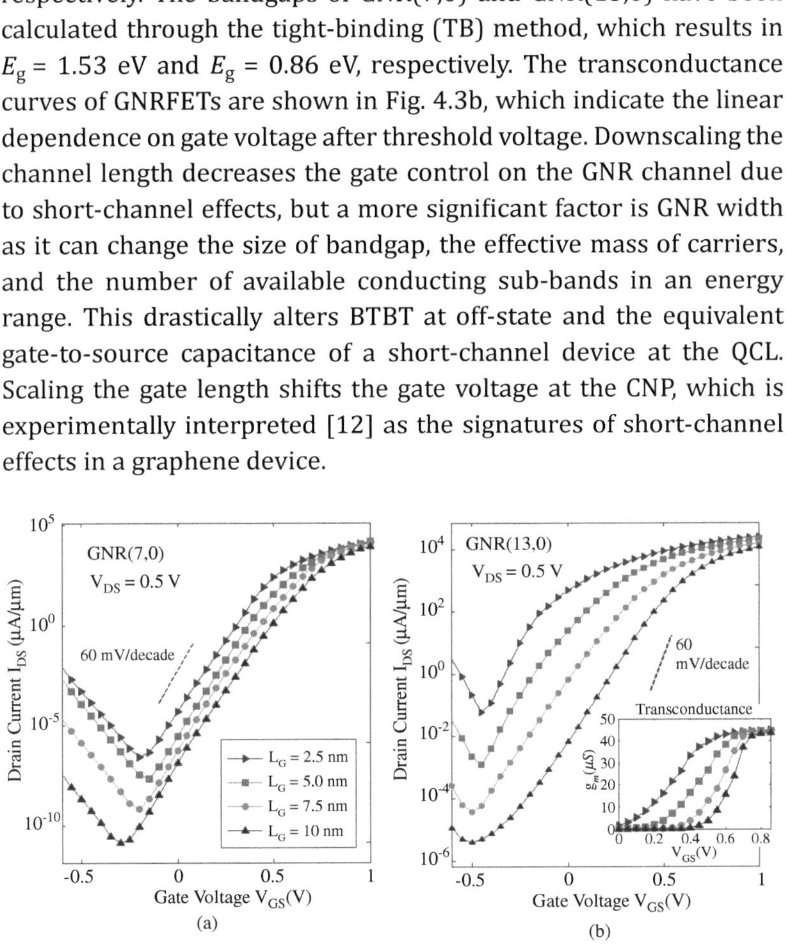

Figure 4.3 Transfer characteristics of (a) GNR(7,0) and (b) GNR(13,0) channels for different channel lengths at $V_{DS} = 0.5$ V. Inside graph shows the corresponding transconductance of GNR(13,0) versus gate voltages. *Note*: Same legends as in (a) are considered for (b). The test device parameters are given in Fig. 4.1. Reprinted from Ref. [6], Copyright 2015, with permission from Elsevier.

4.4 Scaling Effects on Static Metric of GNRFETs

4.4.1 OFF-Current

As the main indicator of low-power design, OFF-current is increased by scaling down the gate length for a given GNR width. Both the height and the width of the channel potential barrier are decreased in the short-channel GNRFETs, which increases both the thermionic emission of carriers passing over the channel barrier and the direct tunneling of carriers through the potential barrier [3]. As it can be seen in Fig. 4.4, the OFF-current per channel width of the FET with GNR(7,0) channel is changed from 2.2×10^{-9} μA/μm to 4.8×10^{-5} μA/μm, and that of GNR(13,0) channel has higher minimum current changing from 2.6×10^{-7} μA/μm to 1.2×10^{-1} μA/μm for scaling the GNR channel length from 15 nm down to 2.5 nm. Therefore, GNR(13,0) not only shows larger OFF-current than GNR(7,0) by scaling the channel length but also has higher increasing trend in OFF-current by scaling the channel length, resulting in GNRFETs with reduced robustness to short-channel effects. For GNRFETs with wider GNRs, such as GNR(13,0), the bandgap is smaller and the carriers have lighter effective mass, which increase the BTBT between the hole states in the channel and the electron states in the drain to some extent, degrading the off-state device performance of wider GNR channels. The OFF-current of narrow GNR channels is promising, compared with the design criteria of silicon-based channels, 100 nA/μm and 10 nA/μm for high-performance and low-power digital ICs.

4.4.2 I_{ON}/I_{OFF} Ratio

Both the ON- and OFF-currents are increased by decreasing the channel length and increasing the GNR width; however, the OFF-current is increased more by tunneling effects, which lead to a significant change in the I_{ON}/I_{OFF} ratio, as shown in Fig. 4.5. For instance, six times shrinking the channel length from 15 nm to 2.5 nm decreases the I_{ON}/I_{OFF} ratio of 15 nm GNR(7,0) from 9×10^{10} to 1.1×10^{8}, approximately three orders of magnitude, while scaling up the channel width approximately twice to GNR(13,0) can deteriorate it more to 6.7×10^7. In an effort to improve the I_{ON}/I_{OFF} ratio, a novel GNRFET structure composed of two side metal gates with smaller work function has been presented [13], which suppresses short-channel effects in GNRFETs by inducing the inversion layers next to the drain and source regions. As can be seen, the I_{ON}/I_{OFF} ratio of GNR(13,0) can only meet the criterion of high-performance design and cannot be a proper channel material for low-power design.

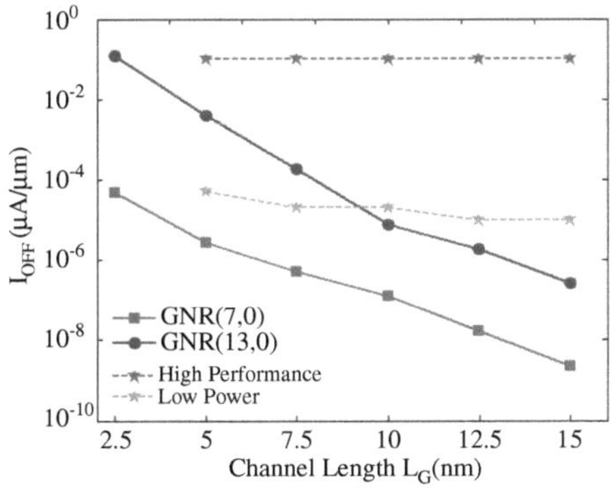

Figure 4.4 OFF-current versus channel length for GNRFETs with channel of GNR(7,0) and GNR(13,0). *Note*: OFF-current of GNRFETs with GNR channel (13,0) has been obtained at V_{DS} = 0.5 V and for V_{GS} close to charge neutrality point, assuming 0.4 eV work function difference between metal gate and graphene. Reprinted from Ref. [6], Copyright 2015, with permission from Elsevier.

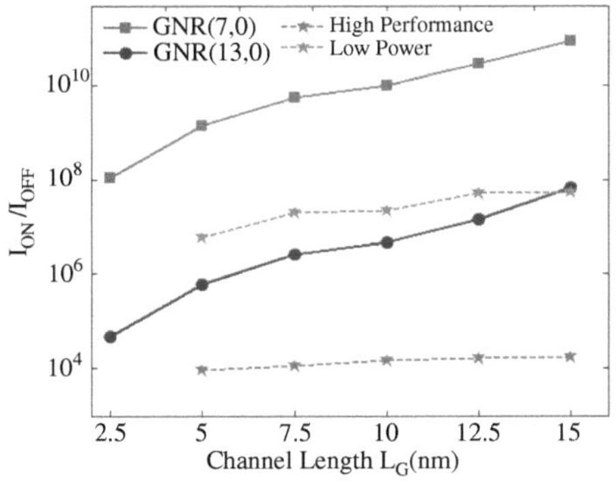

Figure 4.5 I_{ON}/I_{OFF} ratio versus channel length for GNRFETs with channel of GNR(7,0) and GNR(13,0). The I_{ON}/I_{OFF} ratio is obtained referring to the ON-current at V_{GS} = V_{OFF} + 0.8 V. The test device parameters are given in Fig. 4.1. Reprinted from Ref. [6], Copyright 2015, with permission from Elsevier.

4.4.3 Subthreshold Swing

One of the important figures of merit for the standby power dissipation of FETs in ICs is subthreshold swing (SS), which has the fundamental limit of 60 mV/decade at 300 K due to the thermal emission of carriers over the channel potential barrier. In the same scenario as leakage current, the subthreshold slope of GNR(7,0) is sharper than GNR(13,0). Scaling down the channel length from 15 nm to 2.5 nm increases the subthreshold slope of GNR(7,0) from 65 mV/decade to 72 mV/decade, while that of GNR(13,0) increases from 88 mV/decade to 128 mV/decade, as shown in Fig. 4.6. The subthreshold slopes of narrower GNRs, i.e., GNR(7,0), are smaller than 90 mV/decade and 125 mV/decade reported for a 10-nm-scaled Si-MOSFET and a double-gate FinFET, respectively [14]. This indicates the advantage of bandgap engineering in reducing leakage current, together with better gate control on the monolayer GNR channel compared to silicon-based MOSFETs.

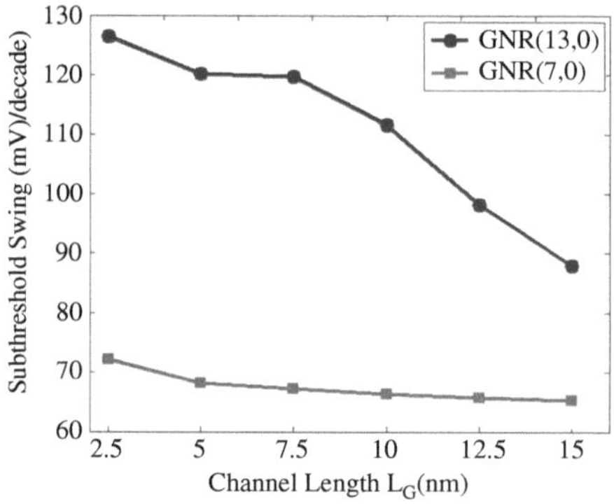

Figure 4.6 Subthreshold swing or GNRFETs with channel of GNR(7,0) and GNR(13,0). *Note*: Subthreshold swing is obtained at V_{DS} = 0.5 V. The test device parameters are given in Fig. 4.1. Reprinted from Ref. [6], Copyright 2015, with permission from Elsevier.

4.4.4 Drain-Induced Barrier Lowering

The short-channel effects degrade the controllability of the gate voltage to drain current, which mainly arises from the lowering of barrier at the beginning of the channel due to the change in drain voltage, known as DIBL. DIBL is a less important performance factor for high-performance logic design [1], but it can be important for low-power IC design. As shown in Fig. 4.7, the DIBL of 15 nm channel length of GNR(7,0) is very small (~7 mV/V), while significantly increasing by channel length scaling to ~200 mV/V for 2.5 nm gate length.

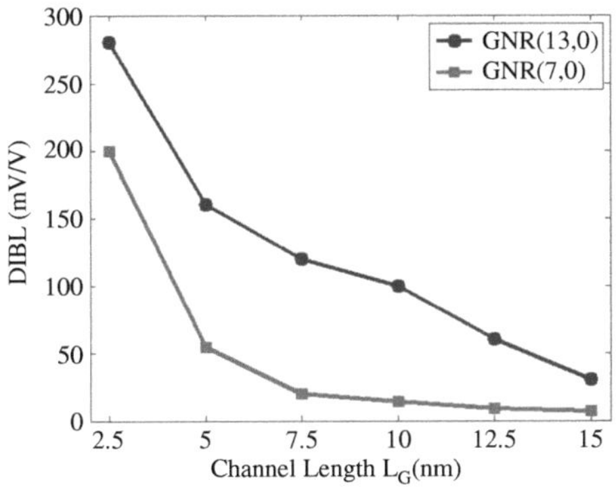

Figure 4.7 DIBL versus channel length for GNRFETs with channel of GNR(7,0) and GNR(13,0). *Note*: DIBL is calculated for the change in the threshold voltage for drain voltages of 0.1 V and 0.5 V. The test device parameters are given in Fig. 4.1. Reprinted from Ref. [6], Copyright 2015, with permission from Elsevier.

The local density of states (LDOS) and current spectrums of the GNR(7,0) channel for two gate lengths of 15 nm and 2.5 nm are shown in Fig. 4.8. It is apparent from the energy-position-resolved LDOS of the device, LDOS(x,E), that the channel potential barrier is decreased by the drain voltage in 2.5 nm gate length, leading to a significant increase in the thermionic emission of carriers passing over the channel barrier and the direct tunneling of carriers through the potential barrier. For a given channel length, the DIBL of the GNR(13,0) channel is larger due to the increase in the contribution of sub-bands in the drain current, which leads to a decrease in the gate electrostatic ability to control the increase in current with an increasing drain voltage at a given gate length.

Further increase in GNR width (smaller bandgap) and the large band bending generated by drain voltage at low gate voltage (smaller gate electrostatic) can increase the BTBT in the drain side

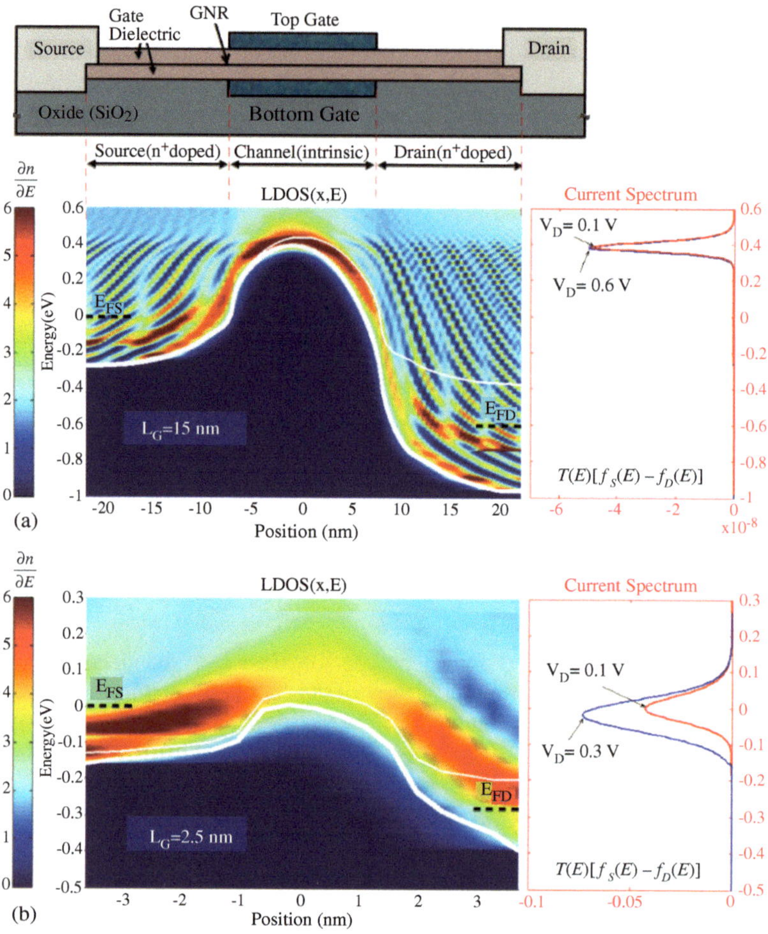

Figure 4.8 Local density of states of GNR(7,0) FET for electrons in the conduction band with the gate lengths equal to (a) 15 nm and (b) 2.5 nm. The first two sub-bands are considered in the transport calculation. The solid lines indicate the band diagram of the first sub-band (conduction band), and the corresponding current spectrums $T(E)[f_s(E) - f_D(E)]$ at two drain voltages of V_{DS} = 0.1 V and 0.6 V (0.3 V) are shown in the figure. *Note*: The color bar shows the number of electrons per unit energy ($\partial n/\partial E$) in correspondence with the DOS. Reprinted from Ref. [6], Copyright 2015, with permission from Elsevier.

of the GNR channel, where the electrons in the valence band of the GNR channel are almost in equilibrium with the Fermi level in the drain region. The phenomena can be observed from the LDOS(x,E)

of GNR(18,0) FET with 10 nm gate length in Fig. 4.9a. In this figure, the bandgap with small LDOS (approximately zero), the source and drain barriers, and the quantum interference pattern due to the incident and reflected waves in the generated quantum well in the valence band of the channel can be easily identified. The I_{DS}–V_{DS} graph is shown in Fig 4.9b, which has no saturation region and not suitable for logic operation due to high output conductance ($g_{ds} = \partial I_{DS}/\partial V_{DS}$). Thus, after the onset of BTBT tunneling (depending on GNR width and bias voltage), the reduction in device performance is not due to the short-channel effects anymore (DIBL and effective channel length modulation) and thereby, there is no benefit for long channel length. Otherwise, this can increase the number of localized states and consequently the positive charge accumulation in the channel, leading to the static feedback and further reduction in the potential energy barrier [15].

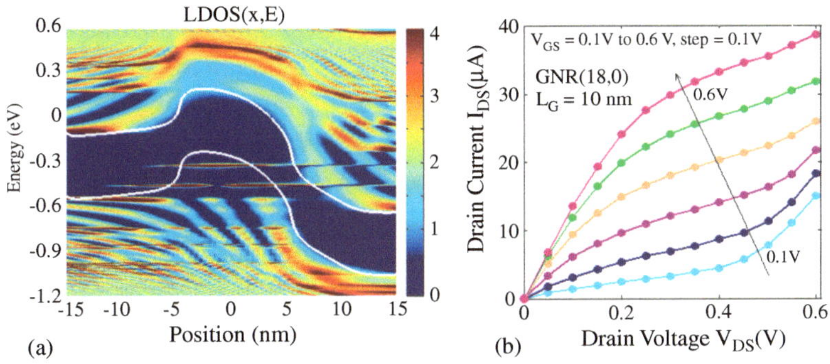

Figure 4.9 (a) Local density of states of GNR(18,0) channel. *Note*: The positions of four sub-bands as well as the conduction and valence bands are shown in figure. (b) I_{DS}–V_{DS} characteristics of GNR(18,0) channel. *Note*: The color bar shows the number of electrons and holes per unit energy. The other parameters of the test device are given in Fig. 4.1. Reprinted from Ref. [6], Copyright 2015, with permission from Elsevier.

4.4.5 Voltage Transfer Characteristic

In addition to the bandgap requirement for low-power design, the complementary operation (normally-off and normally-on devices) is required for digital logic applications. A complementary logic inverter can be designed as one of the main building blocks by integrating two complementary GNRFETs if transistors operate at two sides of their Dirac points [16]. In GNRFETs, the effective masses of electrons and holes are symmetric and thereby the response of pull-up and pull-down networks is equal and opposite, while the asymmetric electron and hole effective mass in conventional silicon CMOS logic needs to be compensated by scaling the physical channel width of the p-type FETs in the pull-up network. Thus, the design of GNRFET logic circuit is easier than that of conventional Si-CMOS circuits; e.g., the switching threshold voltage of a GNRFET-based inverter is in the middle of voltage transfer characteristic (VTC) close to $V_{DD}/2$ [17].

The maximum voltage gain of an inverter (A_{INV}) and noise margin (NM) are two functional criteria of an inverter, which relate to the maximum possible value of a superimposed noise on a digital signal without causing a malfunction of an inversion operation. The maximum voltage gain of an inverter, A_{INV}, can be defined by the maximum slope of VTC in the transition region and the NM can be calculated as $(V_{OH} - V_{IH})/V_{DD}$ where V_{OH} and V_{IH} are the output and input at the unity gain as shown conceptually inside Fig. 4.10a. The VTCs of several GNRFET inverters are shown in Fig. 4.10a. It can be seen that GNR(7,0) with 5 nm gate length exhibits clear voltage inversion with A_{INV} = 4.6 and an ideal rail-to-rail output voltage behavior with NM = 33%V_{DD}. Replacing the channel with GNR(13,0) degrades the A_{INV} and NM of VTC to 4.1 and 29%V_{DD}, respectively, due to the increase in BTBT. Increasing the dielectric constant to κ = 24 (HfO_2) cannot lead to a significant increase in the gate control at the QCL regime; consequently, there is no benefit of using insulator material with larger dielectric permittivity. By shrinking the length of the GNR(13,0) channel to 2.5 nm, the increase in direct tunneling current through the channel potential barrier results in further degradation of A_{INV} and NM to 3.7 and 24%V_{DD}, respectively. By increasing the GNR width in GNRFET-based circuits, the narrow bandgap increases the BTBT leakage current and prevents the pull-down and pull-up networks from completely turning off when its complement network is active. It can be seen that the VTC of GNRFET with the GNR(18,0) channel is significantly deteriorated such that the output voltage swing V_{OS} and gain A_{INV} are decreased to 0.48 V and 1.6, respectively, and the NM regions are nearly diminished.

Figure 4.10b shows the NM degradation of GNRFET inverters by scaling down the supply voltage V_{DD} for the 5 nm channel length of GNR(7,0) and GNR(13,0). It can be seen that GNR(7,0) shows larger NM than GNR(13,0) by scaling V_{DD} such that its NM is above the typical functional criterion of 30% V_{DD} in CMOS logic for scaling down the V_{DD} down to 0.4 V. In the same scenario, GNR(7,0) shows larger maximum inverter gain than GNR(13,0), as shown in Fig. 4.10c. It can be seen that GNR(18,0) has been already deteriorated by BTBT leakage current and its A_{INV} is almost constant close to unity regardless of the value of scaled supply voltage.

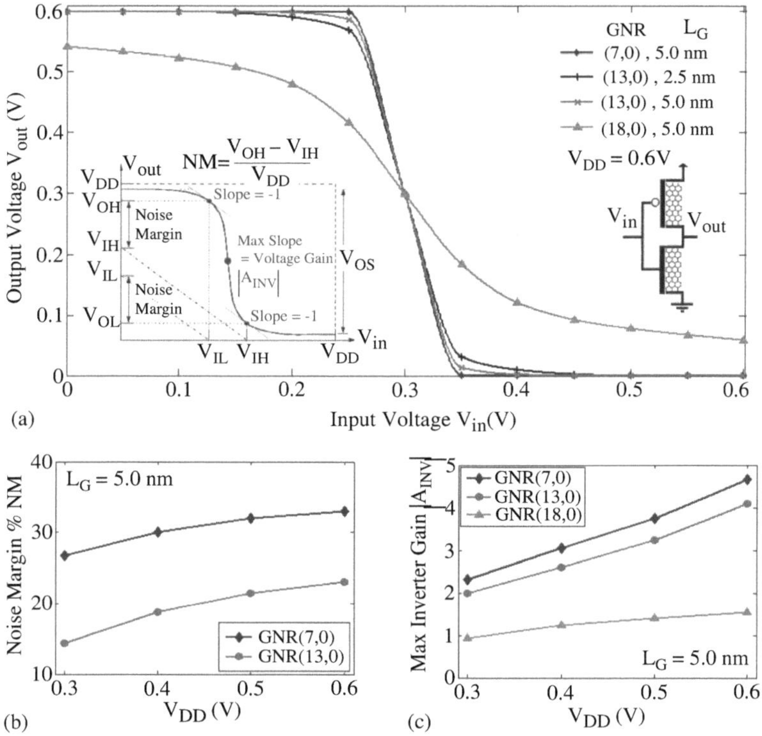

Figure 4.10 (a) Voltage transfer characteristics of a GNR-based inverter for the proposed GNRFET structure with variation of GNR width, gate length, and dielectric constant. *Note*: Charge neutrality point is shifted to $V_{GS} = 0$ by assuming the design with the proper choice of gate work function. Inset conceptually explains the calculation of the noise margin and voltage gain using the VTC of an inverter. (b) Noise margin and (c) maximum gain of GNRFET-based inverters versus scaled supply voltage V_{DD} for the 5 nm channel length of GNR(7,0), GNR(13,0), and GNR(18,0). The other parameters of the test device are given in Fig. 4.1. Reprinted from Ref. [6], Copyright 2015, with permission from Elsevier.

4.5 Scaling Effects on Switching Attributes of GNRFETs

4.5.1 Intrinsic Gate Capacitance

The exceptional properties of graphene, including the mobility, thermal conductivity, and mechanical strength, have resulted in a tremendous research on understanding the switching capabilities of graphene for post-silicon logic applications. The capacitance–voltage (C–V) characteristics are required in order to investigate the GNR intrinsic speed ($I_{DS}/C_G V_{GS}$) as an important speed metric, where C_G is the gate-to-source capacitance. In the QCL, the gate-to-source capacitance is mainly determined by the small density of states of GNR, enforced by the particle-in-a-box boundary condition in the transverse direction, resulting estimation and comparison for the upper-limit performance of GNRFETs. Figures 4.11a and 4.11b show the gate-to-source capacitance versus gate voltage for different channel lengths of GNR(7,0) and GNR(13,0), respectively. It is apparent that the amount of gate-to-source capacitance is decreased by gate length scaling, while its behavior versus gate voltage remains same for all channel lengths. The gate-to-source capacitance becomes very small by approaching the CNP due to small charge in the channel, where the density of states in the energy range created by drain voltage is negligible. It is increased away from the CNP corresponding to its small density of states and maximized after reaching the threshold voltage as most of the higher sub-bands get populated.

It is apparent from Fig. 4.3 that the trend of voltage supply scaling by scaling the channel length is different for GNR(7,0) and GNR(13,0), such that the voltage supply of a wider GNR can be scaled much more than the narrow one. For instance, considering the ON-current of 1.5 μA as the criterion, the gate voltage can scale down from 0.83 V to 0.67 V for scaling the channel length of GNR(7,0) from 10 nm to 2.5 nm, while the gate voltage of the GNR(13,0) channel can reduce from 0.62 V to 0.1 V for the same channel length scaling. This may not be attractive for digital design as the wider GNR has higher leakage current, but it can be used to the advantage of low-voltage design with very-short-channel GNRFETs. It is predicted that increase in current density with the difficulty of scaling V_{DD} results in the enhancement of dynamic power density (CV^2) with channel-length scaling [1], while GNRFETs with short channel length can provide high current density and reach the on-region of operation with small supply voltage together with the other advantages at QCL [18].

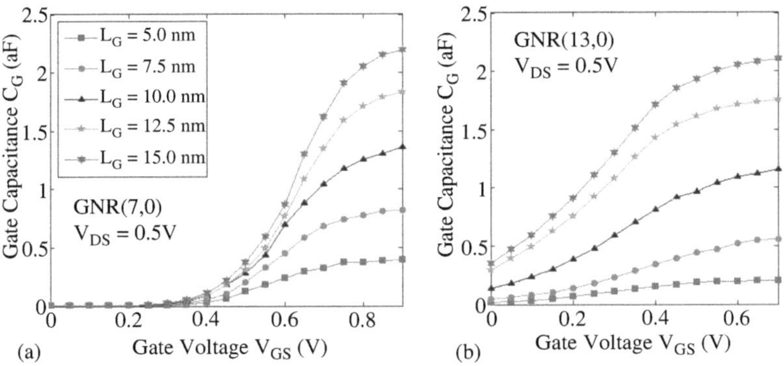

Figure 4.11 Gate capacitance versus gate voltage at V_{DS} = 0.5 V for different channel lengths of (a) GNR(7,0) and (b) GNR(13,0), respectively. The other parameters of the test device are given in Section 4.2. Reprinted from Ref. [6], Copyright 2015, with permission from Elsevier.

4.5.2 Intrinsic Cut-off Frequency

Figure 4.12 shows the intrinsic cut-off frequency $f_T = g_m/(2\pi C_G)$ versus gate voltage for different channel lengths of GNR(7,0) and GNR(13,0), where C_G is the gate-to-source capacitance. The intrinsic cut-off frequency for all channel lengths has reached the terahertz range in the ON-state, but GNR(13,0) has a larger cut-off frequency than GNR(7,0) as the threshold voltage and the impact of density of states are shifted to smaller gate voltages by increasing the GNR width. Decreasing the GNR width opens the bandgap and suppresses the BTBT in GNRFETs, which is achieved at the expense of reducing the electron velocity and degrading the band linearity near Dirac points. The curve inside Fig. 4.10b depicts the intrinsic cut-off frequency of two GNRs versus channel length for the gate voltages of 0.4 V and 0.7 V. It can be seen that for a given channel length below 7.5 nm, the down scaling of voltage supply can be done for the GNR(13,0) channel without a significant drop in the intrinsic cut-off frequency, while the GNR(7,0) channel results in much lower values by the down scaling of the voltage supply with a reduction in the channel length. It should be noticed that the terahertz operation range is due to the assumption of purely ballistic transport, no external series resistance, and negligible parasitic capacitances in order to provide a comparison of the intrinsic upper limit of GNRFET performance, e.g., intrinsic cut-off frequency, intrinsic gate-delay time, and PDP. It is worth mentioning that the terahertz operation of graphene transistor with sub-10 nm gate length has been already demonstrated both theoretically [19] and experimentally [20].

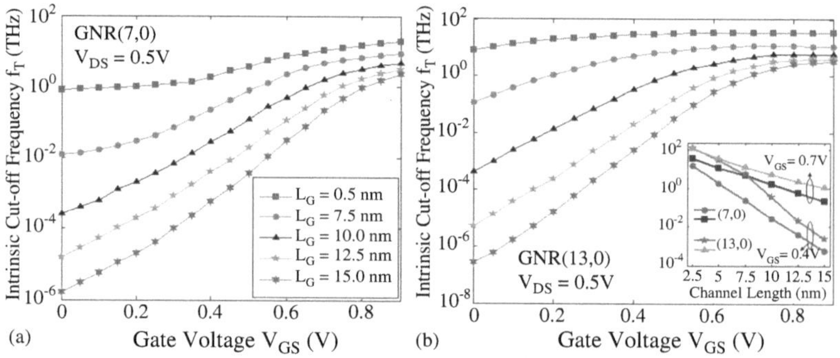

Figure 4.12 Intrinsic cut-off frequency versus gate voltage for different channel lengths of (a) GNR(7,0) and (b) GNR(13,0), respectively. *Note*: The inset shows the intrinsic cut-off frequency of two GNRs versus channel length for the gate voltages of 0.4 V and 0.7 V. The other parameters of the test device are given in Section 4.2. Reprinted from Ref. [6], Copyright 2015, with permission from Elsevier.

4.5.3 Intrinsic Gate-Delay Time

Figure 4.13 shows the intrinsic gate-delay time [1] $\tau = C_G V_{GS}/I_{DS}$ for scaling the channel lengths of GNR(7,0) and GNR(13,0) at three different gate voltages versus the I_{ON}/I_{OFF} ratio for comparison. It is obvious that GNR(13,0) has smaller intrinsic gate-delay time along with smaller I_{ON}/I_{OFF} ratio than GNR(7,0) as upper sub-bands can get highly populated for smaller bandgap and also sub-bands of GNR(13,0) have lighter effective masses and consequently larger carrier injection velocity, which result in higher drive currents at lower supply voltage. The objective is to keep the slope of the intrinsic gate-delay time versus I_{ON}/I_{OFF} ratio as low as possible while scaling down the supply voltage for the sake of decreasing the switching power consumption. Thus, improved transistor operation can be achieved if the difference between the ON- and OFF-currents and the switching speed between these states can be maximized, while the supply voltage can be scaled down at the same time. As shown in the figure by arrows, the slopes of the curves can be kept approximately constant for three scaling transitions of channel length from 10 nm to 7.5 nm, from 7.5 nm to 5 nm, and from 5 nm to 2.5 nm, while the corresponding gate voltages are scaled down from 0.9 V to 0.8 V and then 0.7 V. In other words, when the GNRFET operates at the saturation region, the slopes of both GNR(7,0) and GNR(13,0) are approximately same for all three transitions of channel scaling. However, for a given channel length, the intrinsic gate-delay time of GNR(7,0) is increased more by voltage scaling (VS) than that of GNR(13,0). The ITRS has predicted that such materials can continue improvement in switching speed at the same time with much lower switching power consumption [1]. It can be seen in the figure that both GNR(7,0) and GNR(13,0) can outperform the projection of silicon MOSFETs for low-power and high-performance designs predicted by the ITRS, such that GNR(13,0) would have about 50 times smaller gate-delay time than scaled MOSFETs with 5 nm channel length in 2028.

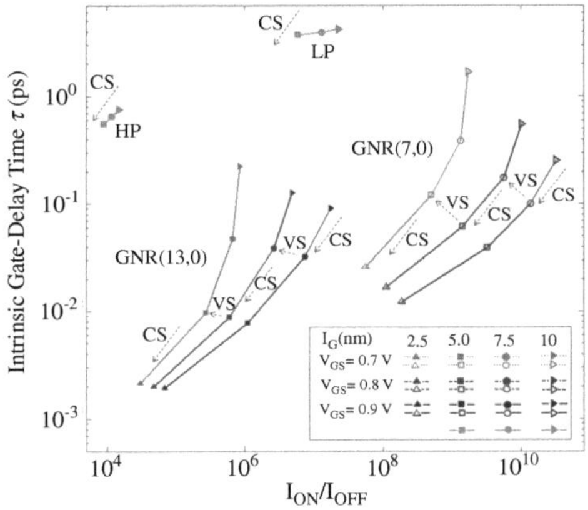

Figure 4.13 Intrinsic gate-delay time versus the I_{ON}/I_{OFF} ratio corresponding to the scaling of the channel length of GNR(7,0) and GNR(13,0) at three different gate voltages. The other parameters of the test device are given in Section 4.1. *Note*: The arrows show the channel scaling (CS) and voltage scaling (VS). The projections of the gate-delay time reported for low-power and high-performance designs by the ITRS are shown as well. CS: channel scaling; VS: voltage scaling; LP: low power; HP: high performance. The other parameters of the test device are given in Section 4.2. Reprinted from Ref. [6], Copyright 2015, with permission from Elsevier.

4.5.4 Power-Delay Product

The energy required for switching a device can be calculated by the PDP $P.\tau = \int Q dV_G$ where Q is the magnitude of charge in the GNR channel. Chin et al. [21] have shown that GNRFET-based logic shows smaller PDP than Si-MOSFET by scaling the channel length as higher carrier velocity of GNR results in higher drive current and thereby smaller delay at the same time with smaller leakage current due to the possibility of bandgap engineering and better control of gate electrostatic on the monolayer GNR channel. Figure 4.14 shows that the PDP is decreased by scaling the channel length for both GNR(13,0) and GNR(7,0), while the static power is increased by scaling the channel length corresponding to the off-current in Fig. 4.4. The trend in reducing PDP by scaling the channel length is more significant at higher gate voltage. GNR(13,0) has smaller PDP at V_{GS} = 0.9 V for all the channel lengths below 15 nm, which remains lower than GNR(7,0) by scaling down both the channel length and supply voltage. The PDP is expected by the ITRS to reduce from the current value of ~0.8 (fJ/µm), reaching ~0.37 (fj/µm) in 2025 for the channel length $L_G \cong$ 7.5 nm and supply voltage $V_{DD} \cong$ 0.7 V. GNR(7,0) and GNR(13,0) show approximately ~0.45 (fJ/µm) and ~0.18 (fJ/µm) for the same channel length and supply voltage. GNR(13,0) has smaller PDP but larger power dissipation for the standby mode due to the higher I_{OFF}, demonstrating better switching behavior.

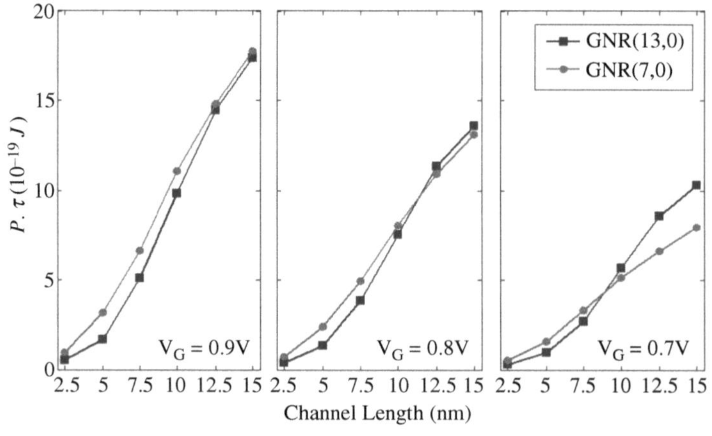

Figure 4.14 Power-delay product versus channel length for GNR(13,0) and GNR(7,0) channels at three different gate voltages. The other parameters of the test device are given in Section 4.2. Reprinted from Ref. [6], Copyright 2015, with permission from Elsevier.

4.6 Summary

In this chapter, we have investigated the static device metrics and switching attributes of GNRFETs for scaling the channel length when the scaling of oxide thickness can no longer result in significant improvement in the GNRFET robustness to short-channel effects. The double-gate GNRFETs have been simulated by solving the quantum transport equation with self-consistent electrostatics in mode space, where the non-parabolic band structure of GNRFET is incorporated by the energy-position-effective mass Hamiltonian. Scaling down the channel length of GNRFETs from 15 nm to 2.5 nm shows that narrow GNRs, e.g., A-GNR(7,0), have superior static performance than wider GNRs, e.g., A-GNR(13,0), and thus decreasing the width of GNRs allows us to compensate the degradation due to the downscaling of the channel, improving the device robustness to short-channel effects. On the contrary, GNRFETs with wider GNR channel show higher ON-state performance by scaling the channel length and supply voltage and improve the switching power and speed for high-performance logic design. By scaling the channel length, the bandgap engineering of GNRFET-based logic circuits provides another degree of freedom for IC designers in order to use GNRFETs with wide and narrow GNR channels for high-performance switching and low-power ICs.

References

1. International Technology Roadmap for Semiconductors (ITRS). (2013). *http://www.itrs.net/.*
2. Yoon, Y., Nikonov, D. E., and Salahuddin, S. (2011). Scaling study of graphene transistors, *2011 11th IEEE International Conference on Nanotechnology*, Portland, OR, pp. 1568–1571.
3. Ouyang, Y., Yoon, Y., and Guo, J. (2007). Scaling behaviors of graphene nanoribbon FETs: A three-dimensional quantum simulation study, *IEEE Transactions on Electron Devices*, **54**, pp. 2223–2231.
4. Fiori, G. and Iannaccone, G. (2013). Multiscale modeling for graphene-based nanoscale transistors, *Proceedings of the IEEE*, **101**, pp. 1653–1669.
5. Banadaki, Y. and Srivastava, A. (2015). Scaling effects on static metrics and switching attributes of graphene nanoribbon FET for emerging

technology, *IEEE Transactions on Emerging Topics in Computing,* **3**, pp. 458–469.

6. Banadaki, Y. M. and Srivastava, A. (2015). Investigation of the width-dependent static characteristics of graphene nanoribbon field effect transistors using non-parabolic quantum-based model, *Solid-State Electronics*, **111**, pp. 80–90.
7. Owlia, H. and Keshavarzi, P. (2014). Investigation of the novel attributes of a double-gate graphene nanoribbon FET with AlN high-κ dielectrics, *Superlattices and Microstructures*, **75**, pp. 613–620.
8. Oh, J. G., Hong, S. K., Kim, C.-K., Bong, J. H., Shin, J., Choi, S.-Y., and Cho, B. J. (2014). High performance graphene field effect transistors on an aluminum nitride substrate with high surface phonon energy, *Applied Physics Letters*, **104**, p. 193112.
9. Konar, A., Fang, T., and Jena, D. (2009). Effect of high-κ dielectrics on charge transport in graphene, *Physical Review B*, **82**, 115452, *arXiv preprint arXiv:0902.0819*.
10. Yoon, Y., Fiori, G., Hong, S., Iannaccone, G., and Guo, J. (2008). Performance comparison of graphene nanoribbon FETs with Schottky contacts and doped reservoirs, *IEEE Transactions on Electron Devices*, **55**, pp. 2314–2323.
11. Lin, Y.-M., Perebeinos, V., Chen, Z., and Avouris, P. (2008). Electrical observation of subband formation in graphene nanoribbons, *Physical Review B*, **78**, p. 161409.
12. Han, S.-J., Chen, Z., Bol, A. A., and Sun, Y. (2011). Channel-length-dependent transport behaviors of graphene field-effect transistors, *IEEE Electron Device Letters*, **32**, pp. 812–814.
13. Mohammadi Banadaki, Y., and Srivastava, A. (2013). A novel graphene nanoribbon field effect transistor for integrated circuit design, *2013 IEEE 56th International Midwest Symposium on Circuits and Systems (MWSCAS)*, Columbus, OH, pp. 924–927.
14. Hasan, S., Wang, J., and Lundstrom, M. (2004). Device design and manufacturing issues for 10 nm-scale MOSFETs: A computational study, *Solid-State Electronics*, **48**, pp. 867–875.
15. Imperiale, I., Bonsignore, S., Gnudi, A., Gnani, E., Reggiani, S., and Baccarani, G. (2010). Computational study of graphene nanoribbon FETs for RF applications, *2010 International Electron Devices Meeting*, San Francisco, CA, pp. 32.3.1–32.3.4.
16. Traversi, F., Russo, V., and Sordan, R. (2009). Integrated complementary graphene inverter, *Applied Physics Letters*, **94**, p. 223312.

17. Tseng, F., Unluer, D., Stan, M. R., and Ghosh, A. W. (2012). Graphene nanoribbons: From chemistry to circuits, In: *Graphene Nanoelectronics: Metrology, Synthesis, Properties and Applications*, H. Raza (Ed.), Springer, Berlin, Heidelberg, pp. 555–586.

18. Knoch, J., Riess, W., and Appenzeller, J. (2008). Outperforming the conventional scaling rules in the quantum-capacitance limit, *IEEE Electron Device Letters*, **29**, pp. 372–374.

19. Sarvari, H., Ghayour, R., and Dastjerdy, E. (2011). Frequency analysis of graphene nanoribbon FET by non-equilibrium Green's function in mode space, *Physica E: Low-Dimensional Systems and Nanostructures*, **43**, pp. 1509–1513.

20. Zheng, J., Wang, L., Quhe, R., Liu, Q., Li, H., Yu, D., Mei, W.-N., Shi, J., Gao, Z., and Lu, J. (2013). Sub-10 nm gate length graphene transistors: Operating at terahertz frequencies with current saturation, *Scientific Reports*, **3**.

21. Chin, H. C., Lim, C. S., Wong, W. S., Danapalasingam, K. A., Arora, V. K., and Tan, M. L. P. (2014). Enhanced device and circuit-level performance benchmarking of graphene nanoribbon field-effect transistor against a nano-MOSFET with interconnects, *Journal of Nanomaterials*, 879813.

Chapter 5

Width-Dependent Performance of GNRFETs

5.1 Introduction

The electronic structure of a graphene nanoribbon (GNR) is very sensitive to the channel width due to its extremely low dimensionality of quasi-one-dimensional (1D) channel. The quantum confinement of a graphene sheet in the form of a 1D nanoribbon with very narrow width (~1–3 nm) provides the energy gap of several hundred meV required for field-effect transistor (FET) operation in digital applications [1, 2]. The precise control of the ribbon width down to the nanometer size is an important technical problem in the experimental characterization of GNRFETs [3] because GNR width can significantly change the bandgap by removing or adding one edge atom along the nanoribbon. Thus, a precise simulation study is required to explore theoretical performance and limitation of GNRFETs for future integrated circuits (ICs).

There is not much reported work on the width-dependent study of GNRFETs with respect to the GNR index. In 2007, Ouyang et al. [4] showed the scaling behavior of GNRFETs considering only one semiconducting family of armchair GNRs. In 2008, Raza and Kau [5] extracted analytical expressions for bandgap and the effective mass of the first sub-band versus GNR width by categorizing them into

Graphene Nanostructures: Modeling, Simulation, and Applications in Electronics and Photonics
Yaser M. Banadaki and Safura Sharifi

ISBN 978-981-4800-36-5 (Hardcover), 978-0-429-02221-0 (eBook)
www.jennystanford.com

three families. In 2011, Sako et al. [6] investigated the effects of edge bond relaxation on device performance using the top-of-the-barrier model for the 10 nm gate length by incorporating the effective mass of the first sub-band. More recently, in 2013, Kliros [7] studied the effect of the width-dependent performance of GNRFETs using an analytical model. However, performance studies on armchair GNR families with channel length below 10 nm have to be researched and a more comprehensive investigation is thus warranted based on more sophisticated approaches.

In this chapter, we provide a comprehensive study on the width-dependent static metrics and switching attributes of two semiconducting families of armchair GNRs (3p,0) and (3p+1,0) focusing on OFF-state current, I_{ON}/I_{OFF} ratio, subthreshold swing, threshold voltage, transfer characteristics, transconductance, gate capacitance, intrinsic cut-off frequency, and intrinsic gate-delay time. We simulate the test device structure in Section 5.2 by solving quantum transport equation with self-consistent electrostatics in mode space. The direct source-to-drain tunneling in short-channel GNRFETs and band-to-band tunneling (BTBT) at the source and drain junctions of wider GNR (small bandgap) can be captured using the proposed quantum transport model. The effect of the non-parabolic band structure of GNRFETs is incorporated using an energy-position-effective mass correction in the quantum transport model (Eq. 3.2), which can be important in determining the subthreshold current, especially by increasing the GNR width as it increases the mismatch between parabolic band and the exact dispersion relation. This discrepancy is important and must be incorporated in the computational model as it can lead to approximately three orders of magnitude underestimation of the leakage current for wider GNRs.

5.2 Device Structure

The double-gate GNRFET structure used in our simulation is shown in Fig. 5.1. In this structure, the GNR is sandwiched between two thin insulator layers in a double metal gate topology to maximize the electrostatic control of the gate electrode over the GNR channel. A hexagonal boron nitride (h-BN) with two-dimensional (2D) atomic structure similar to graphene has been used as a buffer layer [8], which results in a high-k gate insulator free from charge trapping and thereby protection to GNR against environmental influence [9]. As atomic thick graphene is susceptible to environmental conditions of growth, h-BN promotes the growth of a uniform and charge-trapping-free high-k gate insulator [9, 10]. It has large surface optical phonon modes and consequently the lowest remote phonon scattering in thin insulators [11]. This increases the high-temperature and high-electric field performance of graphene on the h-BN substrate.

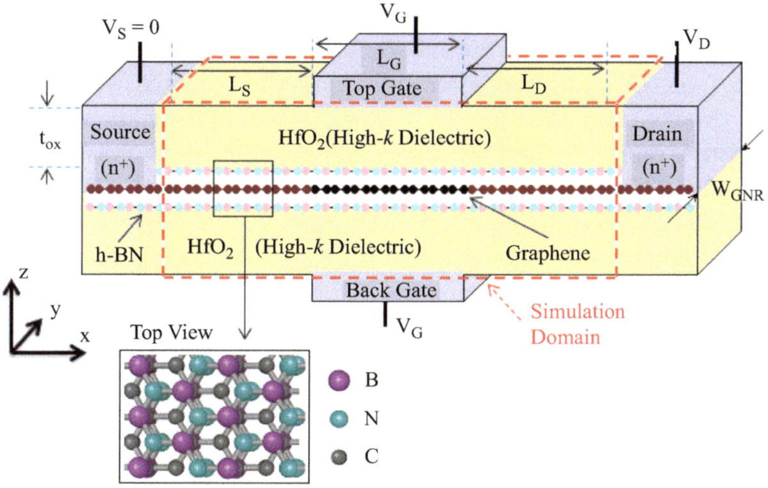

Figure 5.1 Three-dimensional (3D) schematic of the proposed DG GNRFET. *Note*: The armchair GNR channel under the gate area is un-doped, and the source and drain regions are n-type doped. The simulation domain that contains the source, gate, and drain regions in the longitudinal direction is shown with the dashed line. The top view of the GNR sandwiched between two h-BN layers is also shown. Reprinted from Ref. [20], Copyright 2015, with permission from Elsevier.

The proposed GNRFET has the HfO_2 dielectric layer with the relative dielectric permittivity ε_r = 24 and the oxide thickness t_{ox} = 1.2 nm, while the dielectric permittivity of h-BN layers is ε_r = 4 and the interlayer spacing between graphene and h-BN layers is assumed to be 0.3 nm [12]. Thus, the insulator combination of h-BN and HfO_2 dielectrics results in an approximate equivalent silicon oxide thickness (EOT) of 0.5 nm (5 Å), leading to ultimate gate control over the GNR channel [13], which fulfills the criterion of the International Technology Roadmap for Semiconductors (ITRS). In addition, the length of the intrinsic GNR channel, $L_G = 7.5$ nm, and the power supply voltage are based on scaling criteria as in the ITRS for commercial high-performance and low-power FETs for digital ICs. Similar to Section 4.2, the symmetric regions of the GNR channel are heavily doped with the concentration of 0.01 n-type dopants per carbon atom as extensions of source and drain regions and connected to two large metallic contacts.

The h-BN is considered a promising complementary insulator layer for graphene nanostructures [14–17]. It has the same dielectric constant ($\varepsilon \cong 4$) and the breakdown voltage ($V_B = 0.7$ V/nm) as the SiO_2 insulator layer. However, it has smaller bandgap than SiO_2 ($E_{g\text{-hBN}} = 5.9$ eV) and atomically smoother surface with similar lattice constant close to ~1.7%, which is free of dangling bonds and charge traps [18]. First principle methods predict that the difference in interaction energy between the carbon–nitrogen and carbon–boron can open the bandgap of 50 meV [19]; however, there is no experimental evidence of such a bandgap due to the lack of control on crystallographic alignment [10]. We have assumed that the induced bandgap of the h-BN layer is equal to zero in our simulation, unless stated otherwise.

As both the bandgap and the band linearity are altered by GNR width, the importance of non-parabolic correction in static characteristics can be revealed by incorporating the h-BN-induced bandgap in Eq. (3.2) as follows:

$$E_g^b = E_g^{b,\mathrm{GNR}} + \Delta E_{\mathrm{h\text{-}BN}} \tag{5.1}$$

where $E_g^{b,\mathrm{GNR}}$ is the bandgap energy of GNR for a sub-band b and $\Delta E_{\mathrm{h\text{-}BN}} = 50$ meV is the induced bandgap due to the h-BN layer.

5.3 GNR Sub-bands

Figure 5.2 shows the energy of the first four sub-bands at the charge neutrality point (CNP) versus GNR width. It can be observed that the higher sub-bands also follow their own repeating pattern with reduced values in energies by increasing the width of GNRs. The GNR family (3p+1,0) has larger bandgap than its neighbor GNR family (3p,0). The second sub-band of GNRs(3p+1,0) has energy close to the first sub-band energy, which can significantly contribute to carrier transport. Two GNRs(6,0) and (10,0) with the same bandgap E_g = 0.6 eV have been compared in Ref. [6] using a semi-classical model considering only the energy and effective mass of the first sub-band, while we demonstrate that the second sub-band of GNRs(3p+1,0) can contribute largely in carrier transport.

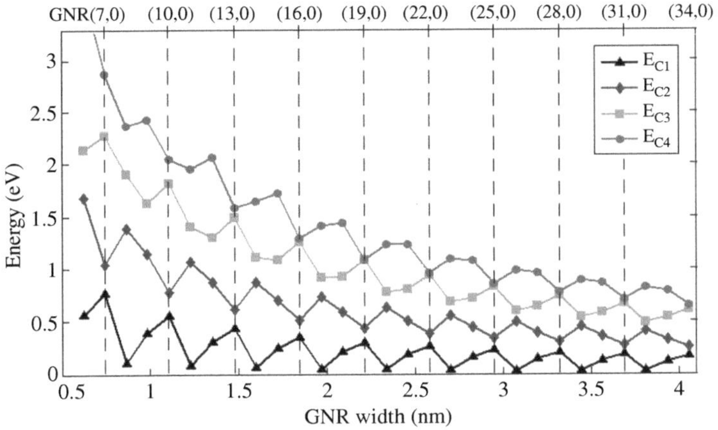

Figure 5.2 Energy of the first four sub-bands at the charge neutrality point versus GNR width. Reprinted from Ref. [20], Copyright 2015, with permission from Elsevier.

The effective masses of the first four sub-bands at the CNP for different members of two GNR groups are shown in Fig. 5.3. The effective masses and energies of the second sub-band can be important for the OFF-state current calculation of the GNR group (3p+1,0) as it can contribute in carrier transport even in low bias condition due to the small energy difference with the first sub-bands. The effective mass of the first sub-band adopted from Ref. [5] demonstrates a close agreement with our results, as shown in Fig. 5.3. It can be observed that the effective mass of the narrow GNR is very sensitive to the width, such that the effective mass of the first sub-band crosses that of the second one and the third sub-band crosses the fourth one for the GNR group (3p+1,0). Likewise the effective mass of the second sub-band crosses that of the third one for the GNR group (3p,0). Thus, an accurate tight-binding (TB) calculation is required to obtain the effective mass and correspondingly correct the non-parabolic band diagram of GNR for width smaller than 3 nm.

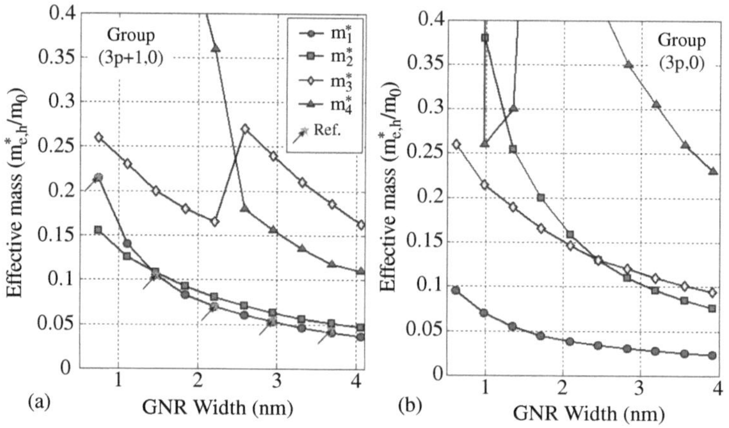

Figure 5.3 Effective mass of the first four sub-bands obtained by TB calculation for (a) GNR group (3p+1,0) and (b) GNR group (3p,0). *Note*: The arrow shows the value from Ref. [5]. Reprinted from Ref. [20], Copyright 2015, with permission from Elsevier.

The drain current as a function of negative and positive voltages at the gate and drain terminals of the proposed GNRFET is shown in Fig. 5.4. For a given drain voltage, a minimum current occurs at the CNP, where the hole current is equal to the electron current and the charge carriers are changed due to the induced electrostatic potential of the gate voltage on sub-bands in the channel. Increasing the drain voltage to positive values leads to the accumulation of holes in the channel due to the increase in BTBT from the source contact to the channel together with the drain-induced barrier-lowering effect in a short-channel device. This increases the minimum current value and shifts it to the positive gate voltage. In a similar scenario, an increase in the drain voltage to negative values can lead to the BTBT from the drain contact to the channel, increasing and shifting the minimum current at the CNP to negative gate voltages. Increasing (decreasing) gate voltage increases the electron (hole) carriers in the GNR channel by shifting the Fermi energy toward the conduction (valence) sub-bands. The ambipolar transport is partially recovered with regard to the subthreshold region created due to the generated bandgap of GNR. This has been already demonstrated experimentally for the GNR with the reduced impurity similar to that of large-area graphene [21].

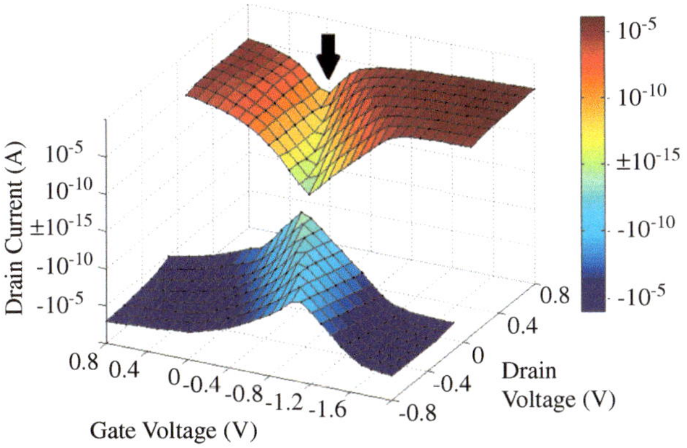

Figure 5.4 Drain current as a function of negative and positive voltages at gate and drain terminals for the GNRFET with the channel of armchair GNR(13,0). *Note*: The channel length, width of metal gate, and gate dielectric thickness are 7.5 nm, 1.48 nm, and 1.2 nm, respectively. The other parameters of the test device are given in Section 5.2. The dielectric constant of the HfO_2 insulator layer and the h-BN buffer layer are 24 and 4, respectively. Arrow indicates passing through the mid-gap energy corresponding to the charge neutrality point. Reprinted from Ref. [20], Copyright 2015, with permission from Elsevier.

5.4 Width-Dependent Static Metrics of GNRFETs

In order to investigate the static characteristics of GNRFETs as a function of GNR width, the armchair GNRs need to be classified into two groups of (3p+1,0) and (3p,0) as their band structure is different and can be altered differently by changing the GNR width. The I_{DS} versus V_{DS} characteristics for two GNR groups of (3p+1,0) and (3p,0) are shown in Fig. 5.5 for the test structure in Fig. 5.1. The strong saturation region for even the short channel length L_G = 7.5 nm indicates good MOSFET-type behavior. On the other hand, the saturation drive current of a typical silicon MOSFET with channel lengths below 10 nm drops due to V_{DD} scaling and significant source–drain tunneling [22]. In addition to short-channel effects, the saturation slope depends on GNR width, such that increasing V_{DS} in a wide GNR can significantly increase the depletion of electrons in the valence band, which corresponds to the accumulation of holes in the GNR channel.

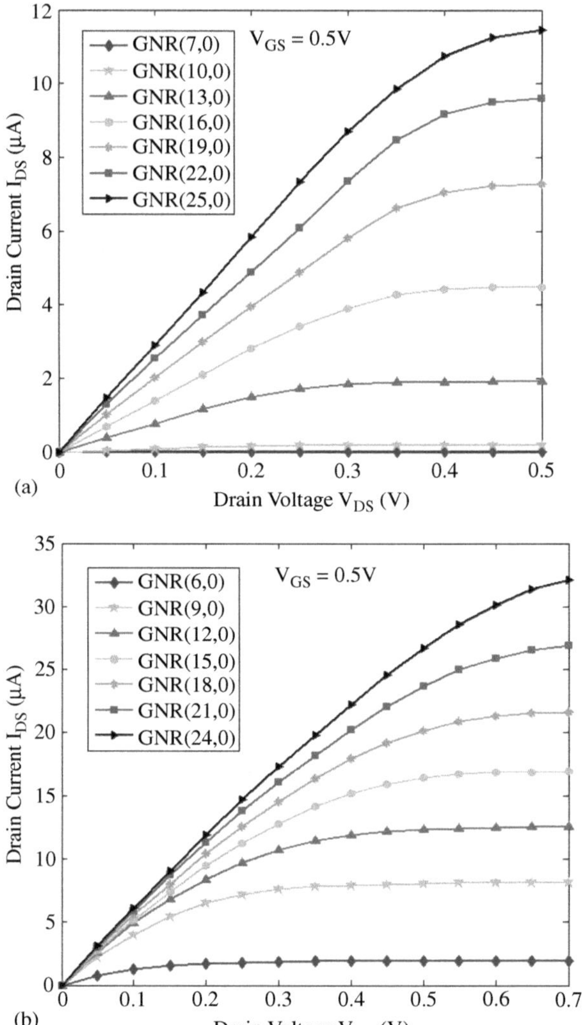

Figure 5.5 Output characteristics for two families (a) GNRs(3p+1,0) and (b) GNRs(3p,0). The parameters of the test device are given in Section 5.2. Reprinted from Ref. [20], Copyright 2015, with permission from Elsevier.

Therefore, the degradation of the subthreshold swing for a short-channel GNRFET with a wide GNR is mostly associated with BTBT and to some extent the direct source-to-drain tunneling, which decreases the dependence of the saturation slope on short-channel effects [23]. As predicted by the ITRS, high mobility and light effective masses of carriers in graphene result in high drive currents at low supply voltage, which can continue the improvement in both the switching speed and low switching power consumption at the same time [22]. Both GNR families can provide approximately an order of magnitude higher drive current than the projected silicon MOSFETs [24]. As it can be observed, GNRs(3p,0) can have about two times higher drive current than GNRs(3p+1,0) because more sub-bands can get populated for smaller bandgap under the same bias condition. In order to increase the drive strength of GNRFETs with narrow GNR, multiple ribbons can be implemented in parallel, which can be connected to two wider contacts [25].

The transfer characteristics I_{DS}–V_{GS} for two GNR groups are shown in Fig. 5.6. For fair comparison between different curves of transfer characteristics, the OFF-voltage of GNRFET, $V_{G,min}$, has been shifted to $V_G \approx 0$ as the CNP of GNRs can be tuned by properly designing the gate work function [4]. In general, increasing GNR width shifts the curve to the smaller gate voltage, leading to smaller threshold voltage for wider GNRs. For the GNR group (3p,0), the drain current is larger, and the threshold voltage can be lower than those of the GNR group (3p+1,0). It can be observed that both ON- and OFF-currents are increased by increasing the width of GNR due to a smaller bandgap and higher number of available conducting sub-bands at a given bias condition.

The first and the narrowest member of the GNR family (3p+1,0) has the OFF-current close to $\sim 2.5 \times 10^{-16}$ A, five orders of magnitude lower than the corresponding first member of the GNR family (3p,0). Significant drain current is due to the thermionic transport of electrons with energies above the potential barrier; however, the smaller bandgap and effective mass in wider GNRs increase the BTBT of the electrons from the channel into the drain region as shown in Fig. 5.7. This BTBT current is still small compared to the thermionic current component. Figure 5.7 shows the energy-position-resolved local density of states of two GNRs (9,0) and (24,0), respectively. The bandgap with quite low local density of states and the source and drain barriers can be easily identified. The quantum interference pattern due to incident and reflected waves in the generated quantum well in the valence band of the channel is also apparent, which has significant contribution at the subthreshold regions for the wide GNRs. This can also increase the leakage current of GNRFETs, making FETs operate like a conductor rather than a transistor.

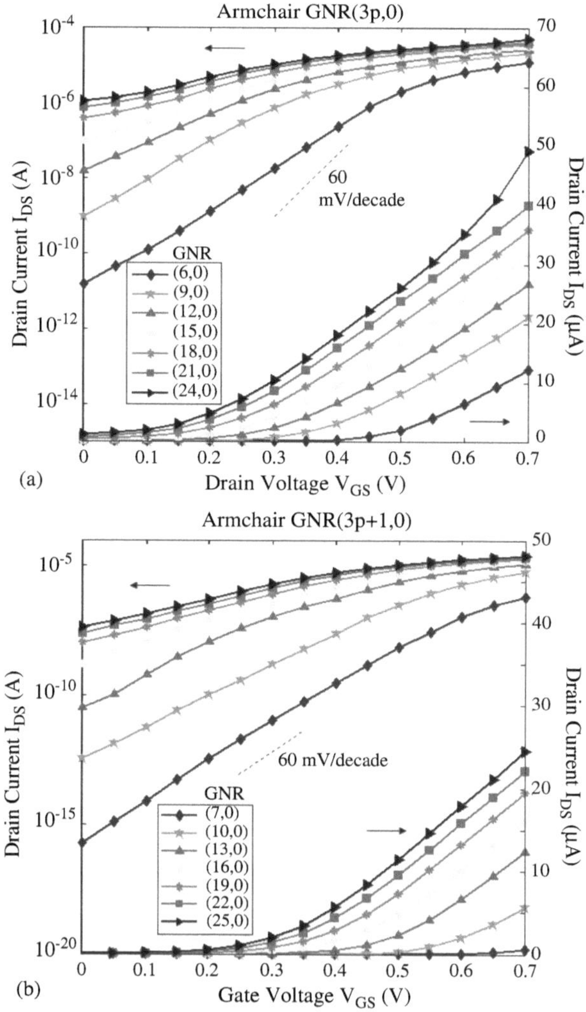

Figure 5.6 Transfer characteristics for two families (a) GNRs(3p,0) and (b) GNRs(3p+1,0) at V_{DS} = 0.5 V. The parameters of the test device are given in Section 5.2. Reprinted from Ref. [20], Copyright 2015, with permission from Elsevier.

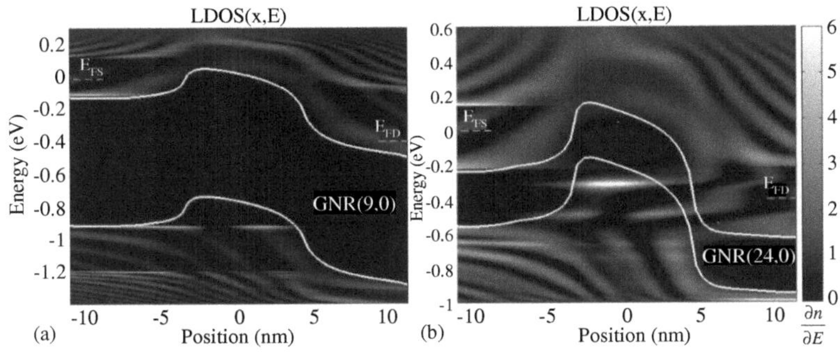

Figure 5.7 Local density of states in (a) GNR(9,0) and (b) GNR(24,0) calculated with the non-parabolic effective mass (NPEM) model considering first two sub-bands at V_{GS} = 0.1 V and V_{DS} = 0.4 V. *Note*: The conduction and valence bands as well as source and drain Fermi levels are shown in the figure. The parameters of the test device are given in Section 5.2. Reprinted from Ref. [20], Copyright 2015, with permission from Elsevier.

5.4.1 OFF-Current

In a typical MOSFET, the OFF-state current is mostly due to the thermionic emission of carriers from over the channel barrier in a longer channel, while in a short-channel device, the decrease in both height and width of the potential barrier in the channel increases the direct tunneling of carriers through the barrier [4]. In GNRFETs, the bandgap and effective mass depend on the type of GNRs, such that the OFF-state current is increased by increasing GNR width, as shown in Fig. 5.8. The OFF-state current of GNRs(3p,0) is larger than GNRs(3p+1,0) as it has smaller bandgap, which provides more available sub-bands to contribute in BTBT from drain contact to channel. In addition, the effective mass and the energy position of upper sub-bands are different for two GNR families, which can significantly change their OFF-state current. As can be seen in Fig. 5.8, GNR(19,0) and GNR(12,0) have different OFF-state characteristics, while they have approximately the same bandgap close to E_g = 0.6 eV. GNR(19,0) has larger effective mass equal to 0.075m$_0$ and 0.085m$_0$ for the first and second sub-bands than that of GNR(12,0) equal to 0.055m$_0$, which results in smaller OFF-state current close to ~4.9 μA/μm compared with ~11 μA/μm for GNR(19,0). This reveals the importance of non-parabolic correction of the GNR band structure, which becomes more important by increasing the GNR width due to increase in band linearity near the CNP, as shown in Fig. 5.8. For a given GNR width, the parabolic assumption leads to smaller OFF-state current, as shown in Fig. 5.8. The difference between the non-parabolic effective mass (NPEM) model and the constant effective mass (CEM) model is increased by increasing the GNR width as the wider GNRs have smaller effective mass and the parabolic assumption can be more erroneous.

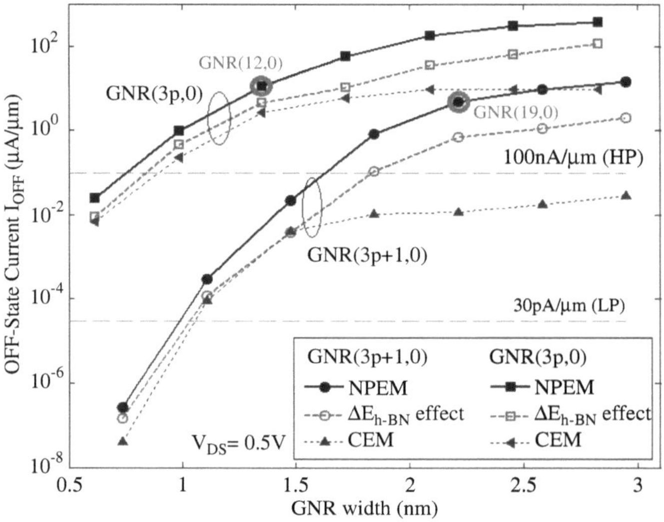

Figure 5.8 OFF-state current of GNRFETs for two GNR families (3p+1,0) and (3p,0) versus GNR width. *Note*: For comparison with the NPEM model, the CEM model is shown with the dotted line. The effect of the induced bandgap of the h-BN insulator layer equal to $\Delta E_{\text{h-BN}}$ = 50 meV is shown with the dashed line. The OFF-current criteria and the I_{ON}/I_{OFF} ratio are also shown. LP: low power and HP: high performance. Two GNRs(12,0) and (19,0) with the same bandgap of 0.6 eV are marked for comparison. The parameters of the test device are given in Section 5.2. Reprinted from Ref. [20], Copyright 2015, with permission from Elsevier.

The OFF-current is one of the design criteria of MOSFETs, as predicted (for 2025 [22]) to be 100 nA/μm and 30 pA/μm for high-performance and low-power digital ICs, respectively. It can be observed that the first three members of GNRs(3p+1,0) and the first member of GNRs(3p,0) have smaller OFF-current than the scaled MOS transistor, promising lower energy consumption of GNRFET-based circuits in the OFF-state. The linearity of GNR energy dispersion is increased by increasing the GNR width, which increases the importance of non-parabolic correction in determining the OFF-state current. In other words, wider GNRs have lighter effective mass of carriers, which can result in higher leakage current, thus requiring non-parabolic correction. The effect of possible induced bandgap of the h-BN layer has been shown to be a function of GNR width. In the same scenario, the bandgap of GNR is decreased by increasing the GNR width, making $\Delta E_{\text{h-BN}}$ an important portion of the equivalent bandgap energy such that the leakage current of wide GNRs is decreased more than narrow GNRs. However, the effect of the non-parabolic band in increasing the OFF-current cannot be possibly reduced by the induced bandgap of the h-BN layer.

5.4.2 I_{ON}/I_{OFF} Ratio

The I_{ON}/I_{OFF} ratio is decreased by increasing the GNR width, following the same trend as the bandgap dependence of GNR width, as shown in Fig. 5.8b. The narrowest ribbon in GNRs(3p+1,0) has the highest I_{ON}/I_{OFF} ratio of $\sim 4.5 \times 10^9$, which is about 350 times and five orders of magnitude larger than the target I_{ON}/I_{OFF} ratio for low-power and high-performance designs, respectively. However, the fifth member of GNRs(3p+1,0) and the third member of GNRs(3p,0) have smaller I_{ON}/I_{OFF} ratios. In an effort to improve the I_{ON}/I_{OFF} ratio, a GNRFET structure composed of two side metal gates with smaller work function has been presented in Ref. [26], which suppresses short-channel effects in GNRFETs by inducing the inversion layers next to the drain and source regions. While the h-BN layer decreases both the OFF-current and ON-current, the I_{ON}/I_{OFF} ratio can be increased due to the increase in the bandgap of GNRs.

5.4.3 Subthreshold Swing

The subthreshold swing is an important OFF-state figure of merit for FETs, which corresponds to the standby power dissipation in ICs. The subthreshold has physical limits and cannot be below $n(k_B T/q)$ $\ln(10) = 60$ mV/dec due to the thermal emission of carriers over the channel barrier, where n is the subthreshold slope factor. While thermionic current is the dominant current for the gate voltage away from the CNP, the BTBT current can strongly contribute to the subthreshold current. It can be observed from Fig. 5.6 that the subthreshold slope is decreased by GNR width as decreasing the bandgap increases the contribution of the BTBT current. Figure 5.10 shows the width dependence of the subthreshold swing for two GNR families (3p+1,0) and (3p,0). As the GNRs(3p+1,0) have larger bandgap than GNRs(3p,0), it demonstrates smaller subthreshold swing such that the range and trend of the subthreshold of the GNR group (3p,0) are more sensitive to width than that of group (3p+1,0). The first member of GNRs(3p,0) has the subthreshold swing equal to 90 mV/decade, while the first member of GNRs(3p+1,0) with subthreshold swing equal to 67 mV/decade can have superior subthreshold performance close to the physical limit of 60 mV/decade for MOS transistors. This value is much smaller than 125 mV/decade and 90 mV/decade reported [27] for a 10 nm scaled double-gate Fin-FET and MOSFET, respectively. It shows the advantage of bandgap engineering of GNRFETs as well as better control of gate electrostatic over atomically thin GNR channel in reducing the subthreshold slope. The range and trend of subthreshold curves indicate that the GNRs in group (3p,0) are more sensitive to width variation than group (3p+1,0).

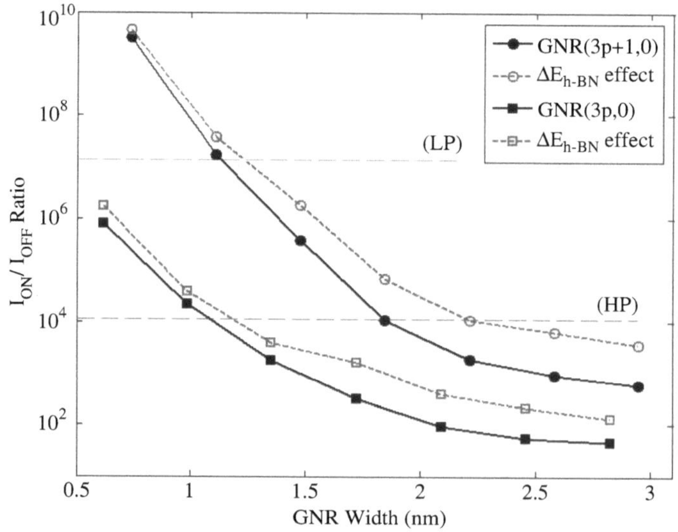

Figure 5.9 The I_{ON}/I_{OFF} ratio of GNRFETs for two GNR families (3p+1,0) and (3p,0) versus GNR width. The effect of the induced bandgap of the h-BN insulator layer equal to $\Delta E_{h\text{-}BN}$ = 50 meV is shown with the dashed line. The I_{ON}/I_{OFF} ratio criteria are also shown (LP: low power and HP: high performance). Reprinted from Ref. [20], Copyright 2015, with permission from Elsevier.

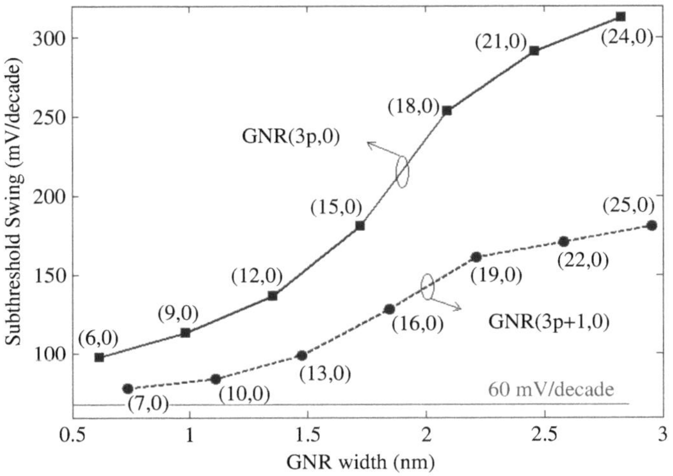

Figure 5.10 Width dependence of the subthreshold swing for two GNR families (3p+1,0) and (3p,0) at V_{DS} = 0.5 V. The parameters of the test device are given in Section 5.2. Reprinted from Ref. [20], Copyright 2015, with permission from Elsevier.

5.5 Width-Dependent Switching Attribute of GNRFETs

In this section, we continue the width-dependent performance of GNRFETs for two semiconducting families of armchair GNRs(3p,0) and (3p+1,0), focusing on its switching attributes such as threshold voltage, transconductance, gate-to-source capacitance, intrinsic cut-off frequency, and intrinsic gate-delay time.

5.5.1 Threshold Voltage

From the transfer characteristics, the threshold voltage of GNRFETs with different widths can be extrapolated, as shown in Fig. 5.11a. The increase in GNR width results in smaller bandgap and thereby decreases the threshold voltage for both GNR families, as shown in Fig. 5.11b. For a GNR with smaller bandgap, the higher number of carriers can be induced in conduction and valence bands by gate potential leading to smaller threshold voltage. In the same scenario, the threshold voltages of GNR(3p,0) are smaller than those of

GNR(3p+1,0) for approximately the same GNR widths as the former has smaller bandgap. For example, the widths of GNR(24,0) and GNR(25,0) are 3.07 nm and 3.19 nm, respectively, with the width difference of only one carbon atom. However, the threshold voltage of GNR(25,0) is approximately 0.3 V, while that of GNR(24,0) is close to 0.2 V. In addition, the drain current of GNR(25,0) at $V_{GS} = 0.4$ V is approximately 5.6 μA, while that of GNR(24,0) is close to 18 μA.

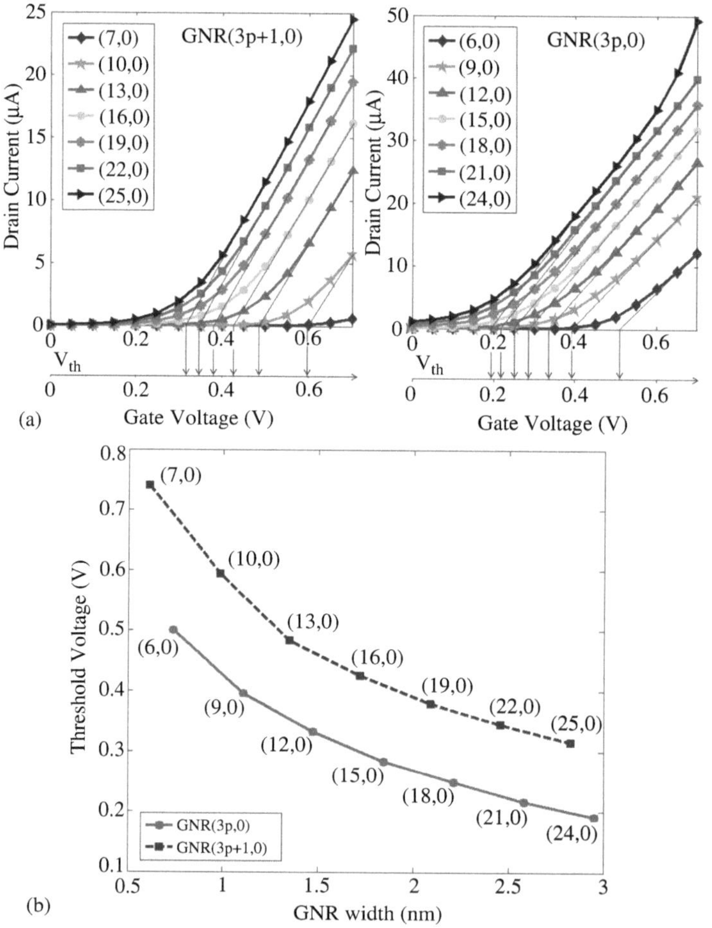

Figure 5.11 (a) Extrapolation of threshold voltages from transfer characteristic. (b) Threshold voltage versus GNR width for two GNR families (3p+1,0) and (3p,0). The parameters of the test device are given in Section 5.2. Reprinted from Ref. [28], Copyright 2016, with permission from IEEE.

5.5.2 Transconductance

The transconductance versus gate bias is shown in Figs. 5.12a and 5.12c. It can be observed that there is a linear dependence on the gate voltage around the threshold voltage, followed by a maximum plateau region. For approximately the same GNR width, GNR(3p,0) has larger transconductance than GNR(3p+1,0) as the former's smaller bandgap results in higher contribution of sub-bands in carrier conduction, showing an inverse trend between transconductance and bandgap. From Figs. 5.12b and 5.12d, it can be observed that the higher transconductance of GNR(3p,0) comes with higher drive current for approximately the same GNR width. For instance, the transconductance and drive current of GNR(24,0) at V_{GS} = 0.45 V are approximately 82 μS and 22 μA, respectively, while these are approximately 62 μS and 8 μA for GNR(25,0).

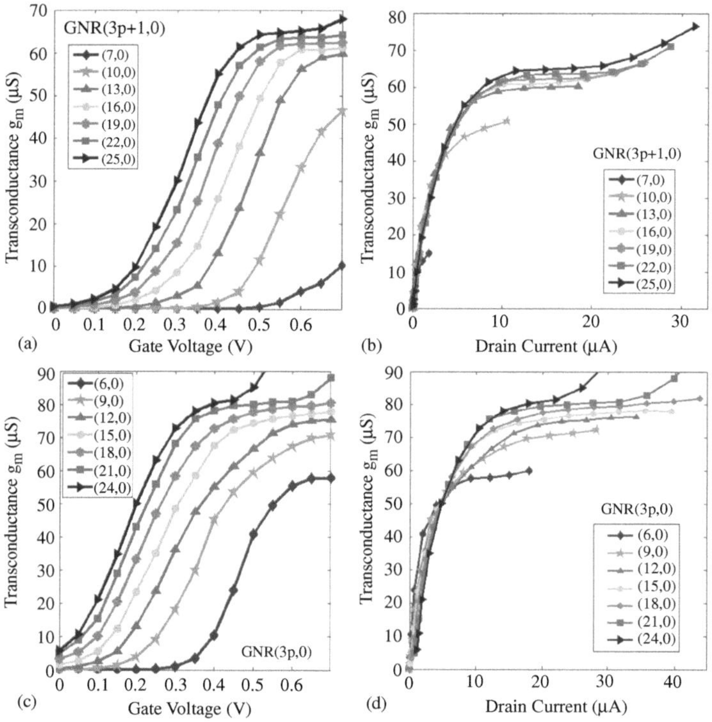

Figure 5.12 Transconductance versus (a) gate voltage, (b) drain current for seven members of GNR(3p+1,0). Transconductance versus (c) gate voltage and (d) drain current for seven members of GNRs(3p,0). The parameters of the test device are given in Section 5.2. Reprinted from Ref. [28], Copyright 2016, with permission from IEEE.

5.5.3 Intrinsic Gate Capacitance

The intrinsic gate-to-source capacitances of GNRFETs in two armchair GNR families are shown in Figs. 5.13a and 5.13d. The gate-to-source capacitance of GNRFETs becomes very small by approaching zero gate voltage, which corresponds to a shift in the Fermi level to the mid-bandgap energy of GNRs and small charge inside the channel. The maximum peak followed by a minimum plateau corresponds to the condition in which the Fermi level passes a peak in the density of state of GNRs. The local maximum of the gate-to-source capacitance of GNRFETs decreases in value and shifts to the smaller gate voltage by increasing the GNR width in both GNR families. To explain the different behaviors of the GNR capacitances versus the gate voltage, the conventional sketch of density of state, similar to Figs. 5.13c and 5.13f, has been converted to a color bar versus the vertical energy axis for two GNR families, as shown in Figs. 5.13b and 5.13e. The blue areas correspond to the bandgap energy of GNRs with very small density of state, while the red areas have the highest density of state (peaks) corresponding to the location of the minimum energies of sub-bands. Shifting the Fermi level in the channel from the mid-energy of bandgap toward higher energies, the first sub-band of GNR(25,0) is the first sub-bands that get populated around E_{F1} = 0.2 eV and result in the corresponding peak in the curve of the gate-to-source capacitance versus gate voltage. As explained in Chapter 3, the quantum capacitance is dominant in the equivalent capacitance of GNRs, and thus the gate-to-source capacitance is related to the derivative of channel charge as follows: $C_G \cong C_Q = \partial Q/\partial V$ ($\propto \partial n/\partial E$). Thus, with regard to the location of the first sub-band in energy, the peak of GNRs with wider bandgaps occurs at a higher gate voltage, such that GNR(7,0) with the largest bandgap in this study has a peak in its gate-to-source capacitance around the gate voltage of 0.8 V corresponding to locating the channel Fermi level around its first sub-band, 0.8 eV (see E_{F2} in Fig. 5.13b). In the same scenario, the behavior of GNR families (3p,0) in Fig. 5.13d can be interpreted using the density of state of GNRs versus energy in Fig. 5.13e. By comparing the peaks and plateaus of the gate-to-source capacitances for two GNR families in Figs. 5.13a and 5.13d, it can be observed that GNR(3p,0) has slightly smaller gate-to-source capacitances than (3p+1,0) for approximately the same GNR width and they also occur at smaller gate voltages. This can be interpreted by comparing the Fermi levels E_{F1} in Figs. 5.13b and 5.13e with regard to the first sub-band of GNR(25,0) and that of GNR(24,0). GNR(25,0) has the second sub-band near the first sub-band that can increase the carrier density in the channel, resulting in higher gate-to-source capacitance for GNR(25,0). These two sub-bands are located at higher energies than the first sub-band of GNR(24,0), which leads to a shift in the behavior of gate-to-source capacitance of GNR(25,0) to the higher gate voltages.

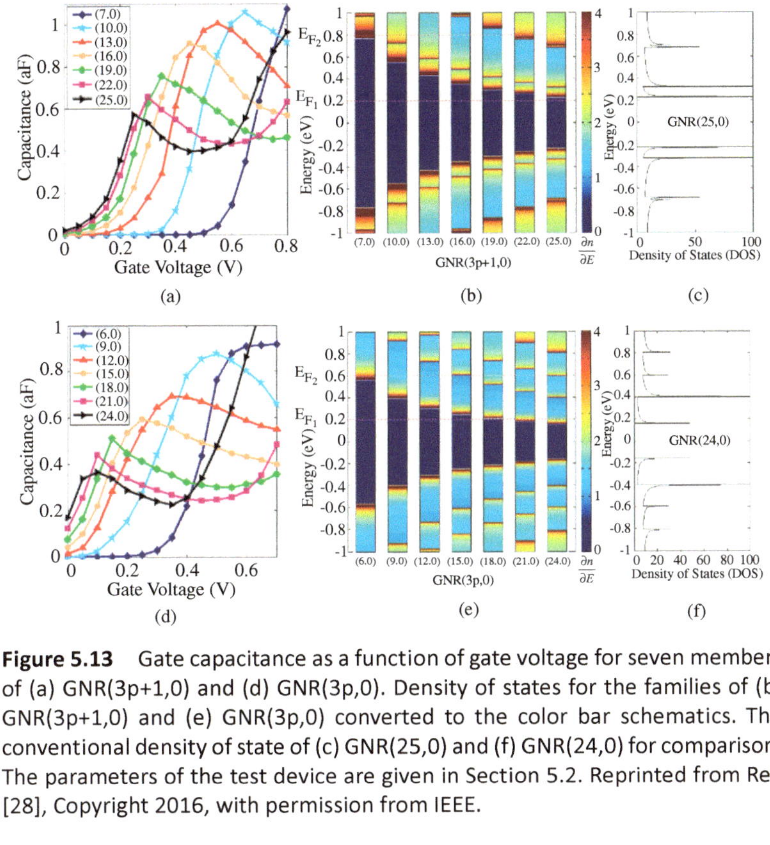

Figure 5.13 Gate capacitance as a function of gate voltage for seven members of (a) GNR(3p+1,0) and (d) GNR(3p,0). Density of states for the families of (b) GNR(3p+1,0) and (e) GNR(3p,0) converted to the color bar schematics. The conventional density of state of (c) GNR(25,0) and (f) GNR(24,0) for comparison. The parameters of the test device are given in Section 5.2. Reprinted from Ref. [28], Copyright 2016, with permission from IEEE.

5.5.4 Intrinsic Cut-off Frequency

The intrinsic cut-off frequency versus gate voltage and drain current of two GNR groups is shown in Fig. 5.14. In general, the wider GNR corresponds to lower bandgap, which leads to observation of higher cut-off frequency at smaller gate bias. It can be observed that

GNR(3p,0) has larger intrinsic cut-off frequency by approximately twice as that of GNR(3p+1,0). The peak of intrinsic cut-off frequency is increased and shifted to lower gate voltages by increasing the GNR width, such that GNR(3p,0) not only has a higher peak of cut-off frequency but also it occurs at a lower gate voltage and a higher drain current. For instance, GNR(25,0) has the cut-off frequency f_T = 25.5 THz at V_{GS} = 0.5 V and I_D = 11.5 μA, while its counterpart GNR(24,0), in another group with approximately the same width, has a higher cut-off frequency f_T = 55 THz at V_{GS} = 0.35 V and I_D = 14 μA. It should be noticed that extremely short channel length L_G = 7.5 nm and the assumption of ballistic transport and negligible parasitic capacitances provide an estimation for the upper limit of the device performance metrics.

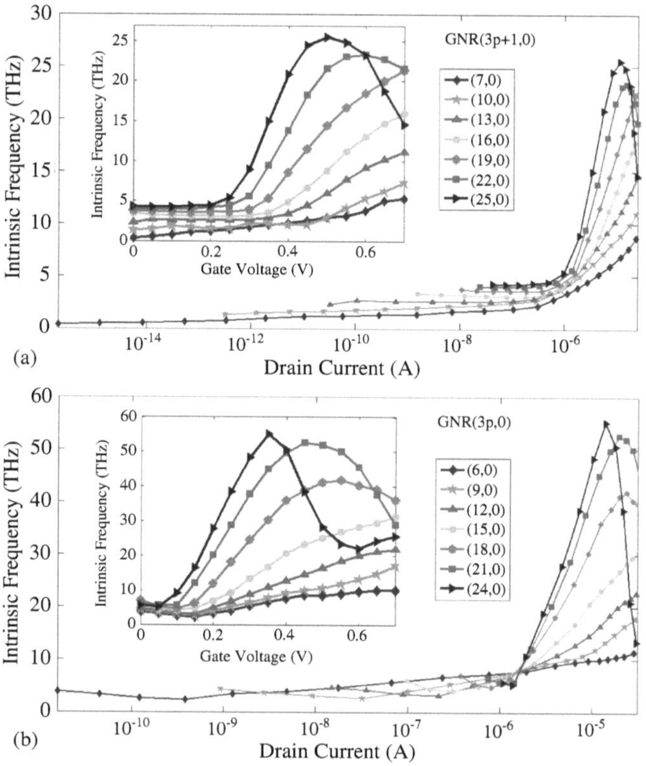

Figure 5.14 Intrinsic cut-off frequency versus drain current for seven members of (a) GNR(3p+1,0) and (b) GNR(3p,0). The inset shows the intrinsic cut-off frequency versus gate voltage. *Note*: V_{DS} = V_{DD}. The parameters of the test device are given in Section 5.2. Reprinted from Ref. [28], Copyright 2016, with permission from IEEE.

5.5.5 Intrinsic Gate-Delay Time

The intrinsic gate-delay time is also another important switching attribute, as shown in Fig. 5.15. The intrinsic gate-delay time is increased by decreasing the GNR width, corresponding to an increase in the I_{ON}/I_{OFF} ratio. GNR(3p,0) has approximately an order of magnitude smaller intrinsic gate-delay time for approximately the same width since their smaller bandgap and effective mass can lead to a more populated upper sub-band and a larger average carrier injection velocity.

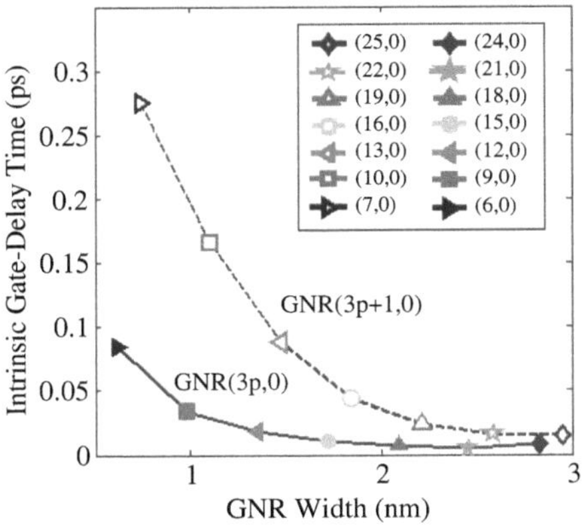

Figure 5.15 Intrinsic gate-delay time versus the I_{ON}/I_{OFF} ratio and GNR width for two families of GNRs(3p+1,0) and GNRs(3p,0) at V_{DS} = 0.5 V and V_{GS} = 0.7 V. The parameters of the test device are given in Section 5.2. Reprinted from Ref. [28], Copyright 2016, with permission from IEEE.

5.6 Summary

The performance of GNRFET is investigated from the computational transport model, as explained in Chapter 3. The model accounts for the tunneling currents on the static performance of GNRFETs in two semiconducting families of armchair GNRs(3p,0) and (3p+1,0). We conclude that increasing the GNR width in both GNR families increases the leakage current and the subthreshold swing and decreases the I_{ON}/I_{OFF} ratio. In this scenario, the GNR group (3p+1,0) leads to superior OFF-state performance such that GNR(7,0) has OFF-state current close to 2.5 × 10^{16} A, five orders of magnitude lower than GNR(6,0) as well as 67 mV/decade subthreshold swing, which is much smaller than that of 90 mV/decade in GNR(6,0). ON-state characteristics such as transfer characteristics, transconductance, gate capacitance, intrinsic cut-off frequency, and intrinsic gate-delay time have also been studied. We found that while the maximum intrinsic cut-off frequencies of both GNR families are increased by increasing the GNR width, GNR(3p,0) shows superior performance such as more than twice larger intrinsic cut-off frequency at lower gate voltages, higher drive current, and lower intrinsic gate-delay time, indicating GNR(3p,0) a more preferable attribute than GNR(3p+1,0) for high-frequency applications.

References

1. Harada, N., Sato, S., and Yokoyama, N. (2013). Theoretical investigation of graphene nanoribbon field-effect transistors designed for digital applications, *Japanese Journal of Applied Physics*, **52**, p. 094301.
2. Johari, Z., Hamid, F., Tan, M. L. P., Ahmadi, M. T., Harun, F., and Ismail, R. (2013). Graphene nanoribbon field effect transistor logic gates performance projection, *Journal of Computational and Theoretical Nanoscience*, **10**, pp. 1164–1170.
3. Cooper, D. R., D'Anjou, B., Ghattamaneni, N., Harack, B., Hilke, M., Horth, A., Majlis, N., Massicotte, M., Vandsburger, L., and Whiteway, E. (2012). Experimental review of graphene, *International Scholarly Research Notices*, **2012**, p. 501686.
4. Ouyang, Y., Yoon, Y., and Guo, J. (2007). Scaling behaviors of graphene nanoribbon FETs: A three-dimensional quantum simulation study, *IEEE Transactions on Electron Devices*, **54**, pp. 2223–2231.
5. Raza, H. and Kan, E. C. (2008). Armchair graphene nanoribbons: Electronic structure and electric-field modulation, *Physical Review B*, **77**, p. 245434.
6. Sako, R., Hosokawa, H., and Tsuchiya, H. (2011). Computational study of edge configuration and quantum confinement effects on graphene nanoribbon transport, *IEEE Electron Device Letters*, **32**, pp. 6–8.
7. Kliros, G. S. (2013). Gate capacitance modeling and width-dependent performance of graphene nanoribbon transistors, *Microelectronic Engineering*, **112**, pp. 220–226.
8. Han, Q., Yan, B., Gao, T., Meng, J., Zhang, Y., Liu, Z., Wu, X., and Yu, D. (2014). Boron nitride film as a buffer layer in deposition of dielectrics on graphene, *Small*, **10**, pp. 2293–2299.

9. Srivastava, A., Banadaki, Y. M., and Fahad, M. S. (2014). Dielectrics for graphene transistors for emerging integrated circuits, *ECS Transactions*, **61**, pp. 351–361.

10. Meric, I., Dean, C. R., Petrone, N., Wang, L., Hone, J., Kim, P., and Shepard, K. L. (2013). Graphene field-effect transistors based on boron–nitride dielectrics, *Proceedings of the IEEE*, **101**, pp. 1609–1619.

11. Ong, Z.-Y. and Fischetti, M. V. (2013). Top oxide thickness dependence of remote phonon and charged impurity scattering in top-gated graphene, *Applied Physics Letters*, **102**, pp. 183506.

12. Fan, Y., Zhao, M., Wang, Z., Zhang, X., and Zhang, H. (2011). Tunable electronic structures of graphene/boron nitride heterobilayers, *Applied Physics Letters*, **98**, p. 083103.

13. Xu, H., Zhang, Z., Wang, Z., Wang, S., Liang, X., and Peng, L.-M. (2011). Quantum capacitance limited vertical scaling of graphene field-effect transistor, *ACS Nano*, **5**, pp. 2340–2347.

14. Dean, C., Young, A., Meric, I., Lee, C., Wang, L., Sorgenfrei, S., Watanabe, K., Taniguchi, T., Kim, P., and Shepard, K. (2010). Boron nitride substrates for high-quality graphene electronics, *Nature Nanotechnology*, **5**, pp. 722–726.

15. Gannett, W., Regan, W., Watanabe, K., Taniguchi, T., Crommie, M., and Zettl, A. (2011). Boron nitride substrates for high mobility chemical vapor deposited graphene, *Applied Physics Letters*, **98**, p. 242105.

16. Levendorf, M. P., Kim, C.-J., Brown, L., Huang, P. Y., Havener, R. W., Muller, D. A., and Park, J. (2012). Graphene and boron nitride lateral heterostructures for atomically thin circuitry, *Nature*, **488**, pp. 627–632.

17. Liu, Z., Song, L., Zhao, S., Huang, J., Ma, L., Zhang, J., Lou, J., and Ajayan, P. M. (2011). Direct growth of graphene/hexagonal boron nitride stacked layers, *Nano Letters*, **11**, pp. 2032–2037.

18. Forster, F., Molina-Sanchez, A., Engels, S., Epping, A., Watanabe, K., Taniguchi, T., Wirtz, L., and Stampfer, C. (2013). Dielectric screening of the Kohn anomaly of graphene on hexagonal boron nitride, *Physical Review B*, **88**, p. 085419.

19. Quhe, R., Zheng, J., Luo, G., Liu, Q., Qin, R., Zhou, J., Yu, D., Nagase, S., Mei, W.-N., and Gao, Z. (2012). Tunable and sizable band gap of single-layer graphene sandwiched between hexagonal boron nitride, *NPG Asia Materials*, **4**, p. e6.

20. Banadaki, Y. M. and Srivastava, A. (2015). Investigation of the width-dependent static characteristics of graphene nanoribbon field effect

transistors using non-parabolic quantum-based model, *Solid-State Electronics*, **111**, pp. 80–90.

21. Lin, Y.-M., Perebeinos, V., Chen, Z., and Avouris, P. (2008). Electrical observation of subband formation in graphene nanoribbons, *Physical Review B*, **78**, p. 161409.
22. International technology roadmap for semiconductors (ITRS). (2013). http://www.itrs.net/.
23. Imperiale, I., Bonsignore, S., Gnudi, A., Gnani, E., Reggiani, S., and Baccarani, G. (2010). Computational study of graphene nanoribbon FETs for RF applications, *2010 International Electron Devices Meeting*, San Francisco, CA, pp. 32.3.1–32.3.4.
24. Wilson, L. (2013). International Technology Roadmap for Semiconductors (ITRS), *Semiconductor Industry Association*.
25. Wang, X. and Dai, H. (2010). Etching and narrowing of graphene from the edges, *Nature Chemistry*, **2**, pp. 661–665.
26. Mohammadi Banadaki, Y. and Srivastava, A. (2013). A novel graphene nanoribbon field effect transistor for integrated circuit design, *2013 IEEE 56th International Midwest Symposium on Circuits and Systems (MWSCAS)*, Columbus, OH, pp. 924–927.
27. Hasan, S., Wang, J., and Lundstrom, M. (2004). Device design and manufacturing issues for 10 nm-scale MOSFETs: A computational study, *Solid-State Electronics*, **48**, pp. 867–875.
28. Banadaki, Y. M. and Srivastava, A. (2016). Width-dependent characteristics of graphene nanoribbon field effect transistor for high frequency applications, *2016 IEEE International Symposium on Nanoelectronic and Information Systems (iNIS)*, Gwalior, pp. 6–10.

Chapter 6

A SPICE Physics-Based Circuit Model of GNRFETs

6.1 Introduction

The exponential trend in scaling metal oxide semiconductor field-effect transistors (MOSFETs) has satisfied Moore's law for decades, leading to denser chips with more functionality, a lower price per chip, faster switching, and lower power consumption. However, the International Technology Roadmap for Semiconductors (ITRS) [1] has predicted the demise of silicon-CMOS (complementary metal oxide semiconductor) technology due to the fundamental limits of CMOSFETs and has put forward alternative channel materials to silicon such as carbon nanotubes and newly discovered graphene [2]. Graphene is a one-atom-thick layer of carbon sheets in a honeycomb lattice, which can outperform state-of-the-art silicon in many applications [2, 3] due to its excellent electronic properties. The carrier transport in graphene is similar to the transport of massless particles since two-dimensional (2D) electron gas in graphene [4] provides both high carrier velocity and high carrier concentration, resulting in a large carrier mobility and, consequently, faster switching capability [5]. Despite the fascinating properties of graphene, it is a semimetal with an overlapping zero bandgap and is not satisfactory for digital applications [6]. The quantum confinement of graphene

Graphene Nanostructures: Modeling, Simulation, and Applications in Electronics and Photonics
Yaser M. Banadaki and Safura Sharifi

ISBN 978-981-4800-36-5 (Hardcover), 978-0-429-02221-0 (eBook)
www.jennystanford.com

sheet in the form of 1D strips with a very narrow width known as graphene nanoribbon (GNR) provides the energy gap of several hundred meV required for FET operations in digital applications [7, 8]. As the fabrication technology of GNRFETs in this structure is still in an early stage, performance evaluation of futuristic graphene-based circuits requires a SPICE-compatible model. The state-of-the-art patterning technique is far from achieving atomic-scale precision, and GNRs with perfect smooth edges cannot be fabricated; such that, line-edge roughness (LER) may play an important role in the production of narrow GNRs for channel material of GNRFETs. The edge roughness enhances the edge scattering and generates edge states in the bandgap, which can significantly enhance the leakage current and reduce the drive current. Thus, modeling edge roughness is very useful to examine the effect of process variation on the circuit performance of GNRFETs. The dispersion of the electrical characteristics due to random edge defects in realistic nanoribbons can be precisely evaluated by statistical analysis at the device level, based on the atomistic quantum transport simulations of large ensembles of randomly generated GNRs [9]. However, the device-level analysis requires extensive computational time; therefore, the same statistical approach cannot be used for circuit-level simulations.

The ideal smooth-edge GNRFETs give an estimation of the upper bound performance; however, the LER needs to be considered for practical GNRFETs, which deteriorate their performance. A semi-analytical model for GNRFETs with perfectly smooth edges was developed in Ref. [10], which involved numerical integrations; thereby, it cannot be used for circuit simulation. In Ref. [11], a circuit model was implemented based on lookup table techniques to use the results of device-level quantum transport simulations for circuit simulations. However, with a single change in a design parameter, the intensive device-level simulations need to be repeated to rebuild the model accordingly, which makes it inappropriate for evaluating the optimized design parameters of GNRFET circuits. A SPICE-compatible model of GNRFETs including the edge roughness is presented in Ref. [12]. In this model, the effect of rough edges on the increasing leakage current of GNR(N,0) is considered by effective bandgap due to the bandgap of GNR($N-1$,0), while the real GNRs with rough edges are composed of all neighboring GNRs. Also the effect of rough edges on decreasing ON-current was modeled by a fitting equation regardless of the physical scattering mechanisms in a GNR channel. In addition, this model cannot capture the effect of large LER on the localization of carriers, which tends to reduce both the OFF- and ON-currents of GNRFETs. It has been shown both experimentally [13] and theoretically [14] that strong localization can appear in single-layer GNRs for high LER.

In this chapter, we develop a physics-based analytical model for circuit simulation of GNRFETs. The band-to-band tunneling (BTBT) from drain to channel regions can be important for small-bandgap GNRs, which has been modeled by a current source in parallel with another current source for the thermionic current. The LER in GNRs is modeled using an exponential autocorrelation function. The model incorporates the effect of edge states on the initial increase in BTBT and high edge scattering of carriers in a localization regime. The device-level simulation is performed to evaluate the static performance of GNRFETs in edge-enhanced BTBT and localization regimes. The results of our analytical model are verified by numerical results from accurate quantum transport simulations based on the non-equilibrium Green's function (NEGF) formalism.

6.2 GNRFET Structure

Figure 6.1 shows the 3D view of a GNRFET, where the ribbon of the armchair chirality GNR is the channel material in a MOSFET-like structure. This structure is expected to demonstrate a higher I_{ON}/I_{OFF} ratio, outperforming the GNRFET with Schottky barriers in logic application [15]. The intrinsic GNR channel (L_{CH}) has the same length underneath as the gate contact (L_G), while its width (W_G) is extended equally from each side of the GNR channel. The width of the intrinsic GNR is $W_{GNR} = (N+1)\sqrt{3}a_{cc}/2$ where a_{cc} is the carbon–carbon bonding length and N is the number of dimer lines for the armchair GNR(N,0). The symmetric regions of the GNR channel between the gate and contacts with the length of L_{RES} are doped with the n-type dopant concentration of f_{dop} per carbon atom as the source and drain reservoirs. The metallic source and drain electrodes are omitted in the model as the two doped regions of GNRs can be directly connected to the GNR interconnect in an all-graphene architecture [16], avoiding the series resistance of metal-to-graphene contacts. Aluminum nitride (AlN) insulator layers with a relative dielectric permittivity of $\kappa = 9$ are assumed. The large-scale and cost-effective production of thin AlN dielectric layers with good reproducibility and uniformity [17, 18] can result in small equivalent oxide thickness (EOT) while reducing phonon scattering in epitaxial graphene, enabling near ballistic carrier transport in a short-channel GNRFET [19].

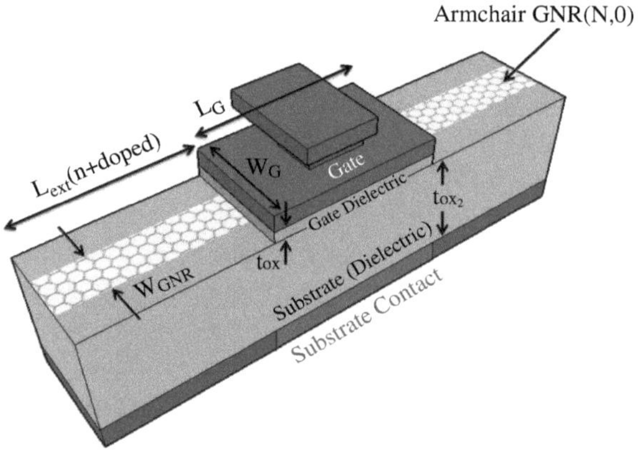

Figure 6.1 3D view of a graphene nanoribbon field-effect transistor (GNRFET) with armchair GNR(*N*,0) as channel material together with the device geometries [61].

6.3 GNRFET Model

In this section, a physics-based analytical model of GNRFET is developed, which allows for GNRFET incorporating the effects of LER as its practical specific non-ideality. The LER is modeled in edge-enhanced BTBT and localization regimes and then verified for various roughness amplitudes.

Figure 6.2a shows the energy band diagram and the corresponding components in the equivalent circuit model of the GNRFET. The model contains four capacitors—$C_{G,CH}$, $C_{S,CH}$, $C_{B,CH}$, $C_{D,CH}$—to account for the electrostatic coupling of the channel to the potentials at gate, source, substrate, and drain electrodes, respectively. Two current sources model the thermal current flowing through the channel and BTBT current from the drain to channel regions. These current sources account for the DC behavior, while a voltage-controlled voltage source (V_{CH}) in the model accounts for charging and discharging the GNR channel, hence the transient AC behavior of GNRFETs.

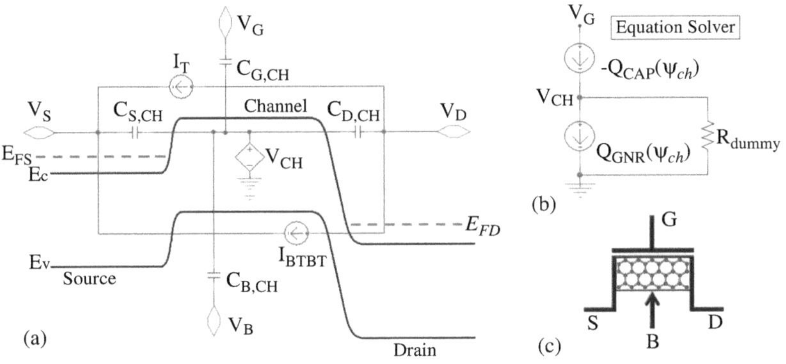

Figure 6.2 (a) Energy band diagram and the corresponding components in the equivalent circuit model of a GNRFET. (b) Series implementation of two voltage-controlled current sources in SPICE to obtain the channel surface potential. (c) GNRFET circuit symbol [61].

6.3.1 Computing GNR Sub-bands

The minimum energy (E_b) and the effective mass (m_b*) of sub-bands for different armchair GNRs need to be obtained for transport equations. Tight-binding (TB) calculation can be employed based on the nearest-neighbor orthogonal p_z orbitals as basic functions equal to the number of atoms in a desired unit cell in the transverse direction [20]. The nearest-neighbor hopping energy for the atoms not located at the edge is $t = -2.7$ eV, while it is assumed $1.12t$ for the pairs of carbon atoms along the edges of the GNR, to take into account the edge bond relaxation due to the lattice termination and occupation of hydrogen atoms at the edges [21]. Further detail about the TB calculation and the effective mass extraction using non-parabolic effective mass model can be followed in Ref. [22].

The bandgap is increased by decreasing the GNR width due to the quantum confinement of carriers in one dimension, while it increases the effective mass due to the degradation of the band linearity near the Dirac point. For narrow armchair GNRs, removing or adding one edge atom along the nanoribbon can significantly change the bandgap energy and effective mass of armchair GNRs. Two-thirds of armchair GNRs, GNR(3p+1,0) and GNR(3p,0), have proper bandgaps, large enough for replacing silicon as a channel material. The third subclass, GNR(3p+2,0), has a very small bandgap. Although the bandgap of GNR(3p+1,0) is slightly larger than that of GNR(3p,0), both GNR families follow the same trends of decreasing bandgaps by increasing the GNR widths. Thus, we have omitted one semiconducting family in our studies to prevent the confusion resulted from chirality dependence of bandgap. Also we have omitted the incorporation of upper sub-bands in the model as the first three sub-bands can accurately describe the carrier transport of GNRFETs, considering the bias voltages and the position of minimum conduction sub-bands in energy.

6.3.2 Finding Channel Surface Potential

The key parameter for evaluating GNRFET current is to find the variation in the channel surface potential (Ψ_{ch}) in response to the variation in gate and source/drain voltages. For a channel material with infinite density of states (DOS), the channel potential can be obtained by geometrical transient capacitance network, such that the channel potential is dominantly controlled by the voltage bias at electrodes, especially gate voltage for long-channel devices. For a semiconducting channel with a finite DOS, the channel surface potential changes with the gate bias at a rate $\Delta\Psi_{ch}/\Delta V_G < 1$. This promotes the device operation close to quantum capacitance limit (QCL) [23]. GNR has very small DOS due to the atomically thin channel in the vertical direction and quantum mechanical confinement in the transverse direction. Thus, evaluating the channel surface potential is very important, which can be modeled using the charge conservation equations by series implementation of two voltage-controlled current sources in SPICE simulation, as shown in Fig. 6.2b [24]. This forces the two currents to be equal in magnitude. In other words, the charge induced by the capacitance networks connected to the contacts (Q_{CAP}) has to be equal to the charge capacity of the GNR channel limited by its DOS. This implementation results in the automatic calculation of the channel voltage (V_{CH}) and the corresponding channel surface potential (Ψ_{ch}).

6.3.2.1 Computing channel charge

In n-type GNRFETs, the hole concentration in the channel is suppressed due to the n-doped drain and source reservoirs; therefore, the electron density of the bth sub-band in the GNR channel (n_b) can be obtained considering the DOS of GNR ($D_b(E)$) from the carrier density relationship as follows [10]:

$$n_b = \int_0^{\infty} f(E) D_b(E) dE \tag{6.1}$$

$$D_b(E) = \frac{2(E_b + E)}{\pi\hbar} \sqrt{\frac{m_b^*}{E_b E(E + 2E_b)}} \tag{6.2}$$

where $\hbar$ is the reduced Planck constant and $f(E)$ is the Fermi–Dirac distribution function. In order to solve the integral in Eq. (6.1) and obtain an analytical equation, the Fermi–Dirac distribution function can be approximated by the Boltzmann distribution $f(E)$ = exp $((E_F - E)/kT$ when the Fermi level is more than $3kT$ away from the sub-band energy [10]. As the bandgap of GNRs can be very small, the assumption is inaccurate in many bias conditions of GNRFETs. To accurately evaluate the electron density in GNRs, $f(E)$ needs to be approximated depending on the relative location of Fermi levels at the terminals to the conduction band energy ($E_{\text{FC}}^b = E_{\text{F},i} - E_{\text{C}}^b$). Equation (6.3) provides a smooth transition between two approximations: (1) exponential carrier concentration (n_b^{exp}) when the Fermi level is near the conduction band (high DOS, $E_{\text{FC}} \cong 0$); and (2) step carrier concentration (n_b^{step}) when the Fermi level is $3kT$ away from the sub-band energy ($E_{\text{FC}} > 3kT$) [12].

$$n_b(E_{\text{FC}}) = w \times n_b^{\text{exp}}(E_{\text{FC}}^b) + (1 - w) \times n_b^{\text{step}}(E_{\text{FC}}^b) \tag{6.3}$$

$$n_b^{\text{exp}}(E_{\text{FC}}) = \frac{\sqrt{m_b^* \alpha^3}\,(1 + 2E_b/\alpha)}{2\pi\hbar E_b} \exp(E_{\text{FC}}/\alpha) \tag{6.4}$$

$$n_b^{\text{step}}(E_{\text{FC}}) = \left(2\sqrt{m_b^*}/\pi\hbar\right) \times \sqrt{\max\left(\left(E_{\text{FC}}(E_{\text{FC}} + 2E_b)/E_b\right), 0\right)} \tag{6.5}$$

where $w = 1/[1+\exp(3(E_{\text{FC}} - kT)/kT)]$ is the relative weight of the two approximations and $\alpha = 3kT/\ln[f(E_{\text{FC}}) \times (1 + \exp((3kT - E_{\text{FC}})/kT))]$. Thus, the total electron density in the GNR channel can be obtained by summation over the carrier density of sub-bands as follows:

$$Q_{\mathrm{GNR}}^{n}=-\frac{qL_{\mathrm{CH}}}{2}\sum_{b}\left[n_b(E-(E_{\mathrm{FS}}-E_{\mathrm{C}}^{b}))+n_b(E-(E_{\mathrm{FD}}-E_{\mathrm{C}}^{b}))\right] \quad (6.6)$$

where q is an electron charge; $E_{\mathrm{C}}^{b}=E_b-\Psi_{\mathrm{ch}}$ is the conduction band energy; $E_{\mathrm{FS}} = E_{\mathrm{F}} - qV_{\mathrm{S}}$ and $E_{\mathrm{FD}} = E_{\mathrm{F}} - qV_{\mathrm{D}}$ are the Fermi levels corresponding to the voltages at source and drain electrodes, respectively. The equilibrium Fermi level of doped reservoirs (E_{F}) sets both E_{FS} and E_{FD} above the conduction band of source and drain regions. We only consider n-type GNRFETs throughout this chapter, though similar analysis can be applied to p-type GNRFETs. In the same scenario, the energy difference between the valence band and the Fermi level at the terminals, $E_{\mathrm{VF}}^{b}=E_{\mathrm{V}}^{b}-E_{\mathrm{F},i}$, and the polarity of the terminal voltage need only to be changed due to the symmetry of the conduction and valence sub-bands in GNRs.

6.3.2.2 Computing transient capacitance charge

The induced charge by capacitance network can be calculated as follows [25]:

$$Q_{\mathrm{CAP}}^{n}=\sum_{i=G,B}C_{i,\mathrm{CH}}\times(V_i-V_{\mathrm{FB},i}-q\Psi_{\mathrm{ch}}) \quad (6.7)$$

where V_{FB} is the flat-band voltage due to the work function difference between metal and graphene. The geometrical capacitances $C_{\mathrm{G,CH}}$ and $C_{\mathrm{B,CH}}$ model the electrostatic coupling between the GNR channel and two electrodes of the gate and the substrate. As the gate width is larger than the GNR width and the oxide thickness, these capacitances can be modeled by the analytical equation of micro-strip lines as follows [26]:

$$C_{i,\mathrm{CH}}=\beta L_{\mathrm{G}}\frac{5.55\times10^{-11}\varepsilon_r}{\ln\left[5.98t_{\mathrm{ox}}/(0.8W_{\mathrm{GNR}}+t_{\mathrm{GNR}})\right]} \quad (6.8)$$

where $t_{\mathrm{GNR}}\cong 0$ is the GNR thickness, t_{ox} is the dielectric thickness, and $\beta = (1 + 1.5\ t_{\mathrm{ox}}/W_{\mathrm{G}})^{-1}$ is a correction term for a case when the gate width is not much larger than the oxide thickness [27].

6.3.3 Current Modeling

Given the surface potential (Ψ_{ch}), both the DC and AC behaviors of GNRFETs can be incorporated in the current calculation. Figure 6.3 shows the energy-position-resolved local density of states (LDOS) of a typical GNRFET, which is numerically simulated by the NEGF formalism [22]. The bandgap with quite low LDOS, potential barrier of the channel together with the source and drain regions, can be easily identified. The quantum interference pattern due to the incident and reflected electron waves in the generated quantum well in the valence band of the channel is also apparent. As can be seen from the LDOS of GNRFETs, the carrier transport can be associated with three mechanisms: (1) thermionic current (I_T) for electrons with energies above the channel potential barrier; (2) direct source-to-drain tunneling current through the channel potential barrier; and (3) BTBT (I_{BTBT}) between the hole states in the source and the electron states in the drain. While the study on the direct source-to-drain tunneling [28] shows that its contribution in the leakage current can be dominant well below the 10 nm channel length, the BTBT can be comparable to the thermionic current at the subthreshold regions of GNRFETs, depending on the bias condition and the bandgap of the GNR channel. We will show that incorporating the two mechanisms of the thermionic current and the BTBT current using two current sources in the equivalent model of GNRFET can result in sufficiently accurate I–V characteristics of GNRFETs for the channel length larger than 10 nm.

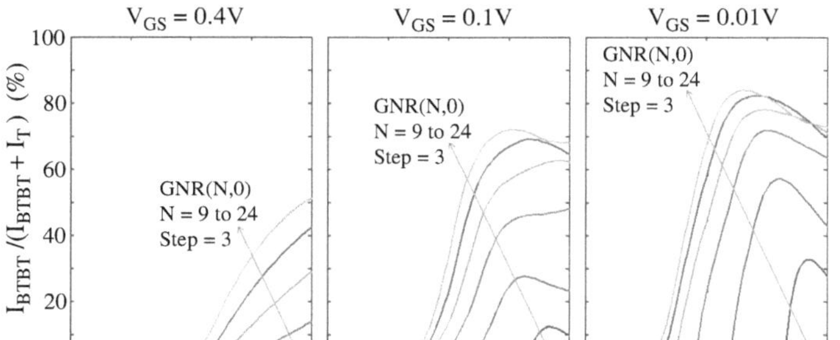

Figure 6.3 Contribution of the BTBT current in the total current of GNRFETs as a function of drain voltages for different gate voltages and various GNR widths [61].

6.3.3.1 Computing thermionic current

The thermionic current can be computed using the Landauer–Buttiker formalism [29], in which the probability of the electrons being injected onto the conduction band from the source side is subtracted from the probability of the electrons being injected onto the conduction band from the drain side as follows:

$$I_{\mathrm{T}} = \frac{2q}{h}\sum_{b}\int_{0}^{\infty} T(E)\left[f(E-(E_{\mathrm{FS}}-E_{\mathrm{C}}^{b}))-f(E-(E_{\mathrm{FD}}-E_{\mathrm{C}}^{b}))\right]dE \quad (6.9)$$

where h is the Planck constant. The integral in the above expression can be evaluated analytically considering the Fermi–Dirac integral of order 0, which results in the current at the ballistic limits as follows:

$$I_{\mathrm{T}} = \frac{2q}{h}k_{\mathrm{B}}T\sum_{b}\Big[\ln(1+\exp((E_{\mathrm{FS}}-E_{\mathrm{C}}^{b})/kT)) - \ln(1+\exp((E_{\mathrm{FD}}-E_{\mathrm{C}}^{b})/kT))\Big] \quad (6.10)$$

6.3.3.2 BTBT current and charge

While the thermionic current strongly dominates the carrier transport at very high drain voltages, the BTBT can be comparable to the thermionic current in the subthreshold region (small V_{GS}) and sufficiently high V_{DS}, as shown in Fig. 6.3. As the wider GNRs can have very small bandgaps and high effective masses, the BTBT can significantly increase the leakage current of GNRFETs. BTBT occurs when the confined states in the valence band of the channel align with the occupied states in the drain. This can happen when the conduction band at the drain side is below the valance band at the channel side ($V_{\mathrm{CH},D} > 2E_b$) and empty states were sufficiently available at the drain side for the electrons that tunnel from the channel region. Assuming ballistic transport for the tunneling process, the BTBT current can be approximated by the maximum possible tunneling current integrating from the conduction band at the drain side up to the valance band at the source side times the BTBT probability as follows [24]:

$$I_{\mathrm{BTBT}} = \frac{2q}{h} k_{\mathrm{B}} T \sum_{b} \left[T_{\mathrm{BTBT}} \ln\left(\frac{1+\exp((qV_{\mathrm{CH},D} - E_b - E_{\mathrm{F}})/k_{\mathrm{B}}T)}{1+\exp((E_b - E_{\mathrm{F}})/k_{\mathrm{B}}T)} \right) \times \frac{\max(qV_{\mathrm{CH},D} - 2E_b, 0)}{qV_{\mathrm{CH},D} - 2E_b} \right] \tag{6.11}$$

where E_{F} is the Fermi level of the doped regions at the drain side of GNRFET, and E_{BTBT} is the Wentzel–Kramers–Brillouin-like transmission coefficient, which can be calculated following the work of Kane [30] as follows:

$$T_{\mathrm{BTBT}} \approx \frac{\pi^2}{9} \exp\left(-\frac{\pi m_b^{*(1/2)} (\eta_b 2E_b)^{3/2}}{2^{3/2} q\hbar F} \right) \tag{6.12}$$

In Eq. (6.12), $F = (V_{\mathrm{CH,D}} + (E_{\mathrm{F}} - \Psi_{\mathrm{ch}})/q)/l_{\mathrm{relax}}$ is the electrical field triggering the tunneling process through the junction at the drain side of the GNR channel when the potential across the drain–channel junction is $V_{\mathrm{CH,D}}$. η_b models the bandgap-narrowing effect under a high electrical field [31], which is set to 0.5 corresponding to that of carbon nanotube in Ref. [24]. Basically, a graphene nanoribbon channel is an unfolded lattice structure of carbon nanotube with the same 1D channel. As such, it is an appropriate assumption that both have the same bandgap-narrowing effects under a high electrical field. l_{relax} is the relaxation length of potential drop, which has been extracted with the procedure explained later in this chapter.

Although the thermionic emission of holes into the channel is negligible for n-type GNRFET, the BTBT significantly increases the accumulation of holes in the channel, especially at a high V_{DS}. As such, both the charges of the GNR channel (Eq. (6.6)) and the charge induced by the capacitance network (Eq. (6.7)) need to be corrected corresponding to the tunneling coefficient (Tr) as follows:

$$Q_{\mathrm{GNR}} = Q_{\mathrm{GNR}}^{n} + Tr.p_b(E_{\mathrm{V}}^{b} - E_{\mathrm{FD}}) \tag{6.13}$$

$$Q_{\mathrm{CAP}} = Q_{\mathrm{CAP}}^{n} + Tr.\beta.C_{i,\mathrm{CH}} \times ((E_{\mathrm{V}}^{b} - E_{\mathrm{FD}})/q) \tag{6.14}$$

$$Tr = 1 - \left[1 + \exp\left(\frac{qV_{\mathrm{CH},D} - \eta_b E_b - E_{\mathrm{F}}}{\delta} \right) \right]^{-1} \tag{6.15}$$

where $E_{\mathrm{V}}^{b} = -E_b - \Psi_{\mathrm{ch}}$ is the valence band energy and $\beta = l_{\mathrm{relax}}/L_{\mathrm{CH}}$. δ is a fitting parameter, which controls how fast Tr increases by increasing the band bending between the channel and the drain, corresponding to the value of $V_{CH,D}$. The transient capacitance network in Fig. 6.2b can be computed by introducing the intrinsic capacitors as the derivatives of the channel charge with respect to drain and source voltages, $C_{S,CH} = \partial Q_{GNR}/\partial V_S$ and $C_{D,CH} = \partial Q_{GNR}/\partial V_D$, which can be implemented in SPICE by voltage-controlled capacitors.

6.3.4 Non-ballistic Transport

The experimental results show that the carrier mobility in graphene can be as high as 200,000 cm^2/Vs [32]. However, different scattering mechanisms due to the intrinsic acoustic phonons (APs) and optical phonons (OPs) of graphene [33], the interaction of carriers with OPs of the substrate [34] and the LER in narrow GNRs [35] can limit its mobility to orders of magnitude lower values. While the transmission coefficient of the carriers can be assumed to be at unity for developing the compact model based on ballistic assumptions [25], these scattering mechanisms must be incorporated in the model as they have been shown to play an important role in the performance of GNRFETs [36]. The transmission of carriers is decreased by scattering in the channel, which can cause a carrier to return to the source region under a low drain bias. The backscattering of carriers to the source continues under a high drain bias within an approximate critical length of $l = (\hbar\omega_{op}/qV_D)L_{CH}$ near the source end of the channel, while the scattered carriers in the channel will be absorbed by the drain beyond this critical distance without having a direct effect on the source–drain current [36]. In the absence of scattering mechanisms, the carrier transport is in the ballistic regime and the conductance is independent of the device length. However, carrier scattering in the channel results in the diffusive transport of carriers, thereby making the conductance inversely proportional to the channel length. The channel transmission coefficient provides a simple way to describe the device in the presence of scattering mechanisms [37] as follows:

$$T = \begin{cases} \lambda^{eff}/(\lambda^{eff} + L_{CH}) & \text{if } qV_D < \hbar\omega_{op} \\ \dfrac{\lambda^{eff}}{(\lambda^{eff} + (\hbar\omega_{op}/qV_D)L_{CH})} & \text{if } qV_D > \hbar\omega_{op} \end{cases} \tag{6.16}$$

where $L_{CH}(=L_G)$ is the channel length and $\hbar\omega_{op} \cong 0.18$ eV [38] is the OP energy and λ^{eff} is the effective mean free path (MFP) of GNRs, which can be obtained using Mattheissen's rule as follows [39]:

$$\frac{1}{\lambda^{eff}} = \frac{1}{\lambda^{sub}} + \frac{1}{\lambda^{ac}} + \frac{1}{\lambda^{LER}} \tag{6.17}$$

where λ^{sub} is the substrate-limited MFP, which is reported close to 100 nm and 300 nm for the GNR on top of SiO_2 and h-BN dielectrics [39], respectively. λ^{ac} is the AP-limited MFP [34] as follows:

$$\lambda_{ap} = \frac{h^2 \rho_s v_s^2 v_f^2 W_{GNR}}{\pi^2 D_A^2 k_B T} \tag{6.18}$$

where $v_s = 2.1 \times 10^4$ m/s is the sound velocity in graphene, $D_A = 17 \pm 1$ eV is the acoustic deformation potential, and $\rho_s = 6.5 \times 10^{-7}$ kg/m^2 is the 2D mass density of graphene. λ^{LER} is the LER scattering-limited MFP, which can be as small as a few tens of nanometers [40], which can exhibit the dominant scattering mechanism in narrow GNRs as it has been predicted in both experimental data [35] and theoretical studies [41, 42]. The edge disorder has been analytically modeled by Anderson distribution in Ref. [43] assuming that atoms at the edges are randomly removed with uniform probability, such that the correlation between edge disorders has been neglected. As LER is a statistical phenomenon, a more realistic model needs an autocorrelation function as have been already used for modeling Si/SiO_2 interface roughness [44] and the LER in GNRs [45, 46]. In this chapter, we consider an exponential spatial autocorrelation function as follows:

$$R(x) = \Delta W^2 \exp\left(-\frac{|x|}{\Delta L}\right) \tag{6.19}$$

where ΔW is the root mean square of the width fluctuation amplitude or roughness amplitude and ΔL is the roughness correlation. Increasing ΔW or decreasing ΔL initially makes the carrier transport more diffusive. The LER causes fluctuations in the edge potential and bandgap modulation due to the localized edge states, as shown in Fig. 6.4a. The decrease in the conductivity of narrow GNRs as a result of LER can be incorporated by introducing the effective bandgap in the transport calculation corresponding to the LER scattering-limited MFP as follows [46]:

$$\frac{1}{\lambda^{\mathrm{LER}}}=\sum_{b}\frac{1}{\lambda_{b}^{\mathrm{LER}}}=\sum_{b}\frac{1}{A\{(E-E^{b})+B(E-E^{b})^{2}\}} \tag{6.20}$$

$$\Delta E_{\mathrm{g}}=2\left(\frac{2k_{\mathrm{B}}T\,L_{\mathrm{CH}}}{AB}\right)^{1/3} \tag{6.21}$$

$$A=\left(\frac{W_{\mathrm{GNR}}}{\Delta W}\right)^{2}\frac{\hbar^{2}}{8m_{b}^{*}\Delta L E_{b}^{2}},\ B=\frac{8m_{b}^{*}\Delta L^{2}}{\hbar^{2}} \tag{6.22}$$

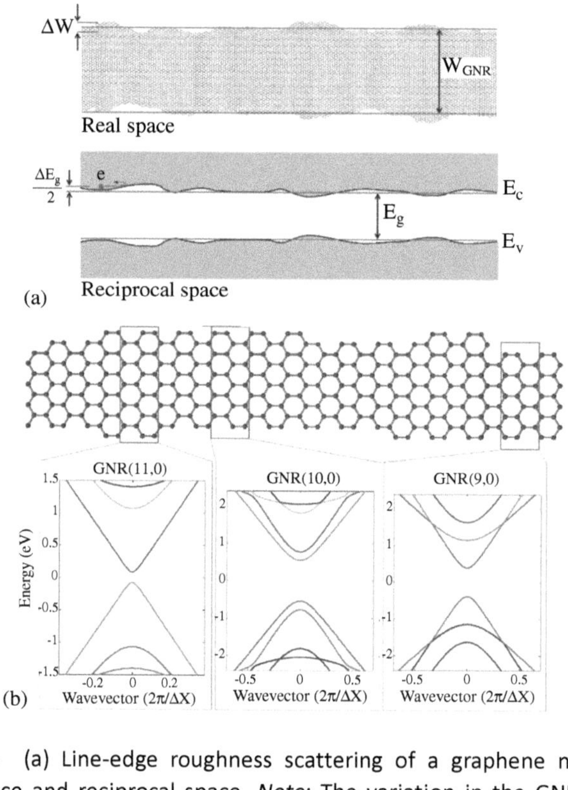

Figure 6.4 (a) Line-edge roughness scattering of a graphene nanoribbon in real space and reciprocal space. *Note*: The variation in the GNR width in real space causes the generation of edge states and potential variation of its bandgap in reciprocal space. (b) Removing or adding carbon atoms at the edge (e.g., GNR(10,0)) can significantly change the local bandgap of GNR and the corresponding change in the local states can contribute in the enhancement of the BTBT current at small roughness amplitude [61].

Using Eq. (6.21), the effective sub-band energy of GNR(N,0), $E_{b,\text{eff}}^{N}$, can be modeled as follows:

$$E_{b,\text{eff}}^{N} = E_b^N + \gamma(\Delta E_g/2)(E_b^N/E_1^N) \tag{6.23}$$

where γ is a fitting parameter, which weighs the increase in the sub-bands energy corresponding to the calculated effective bandgap due to the LER scattering mechanism. While the increase in the sub-band energy of GNRs in Eq. (6.23) can model the decrease in carrier transport, the LER can also contribute to the formation of some localized states in the bandgap, which enhances the BTBT of carriers, leading to the initial increase in the OFF-current of GNRFETs at small roughness amplitudes [47]. As a GNR channel has a large number of carbon rings and is long enough to provide sufficient averaging, the increase in BTBT current of GNR(N,0) due to the variation in its width can be analytically modeled by summation over the BTBT of its neighbor GNRs (conceptual examples of N = 9, 10, and 11 are shown in Fig. 6.4b as follows:

$$I_{\text{BTBT,rough}}^{N} = I_{\text{BTBT}}^{N} + \alpha_r \frac{\sum_{i=1}^{m}\left[\left(I_{\text{BTBT}}^{N-i} + I_{\text{BTBT}}^{N+i}\right)\left(\frac{\Delta W}{W_{\text{GNR}}} - \frac{i}{N-1}\right)^{-1}\right]}{\sum_{i=1}^{m}\left[\frac{\Delta W}{W_{\text{GNR}}} - \frac{i}{N-1}\right]^{-1}} \times \exp\left(-\left(\left(\frac{\Delta W}{W_{\text{GNR}}} - \frac{\Delta W_c}{W_{\text{GNR}}}\right)\Big/\beta_r\right)^2 + \frac{N}{2}\right) \tag{6.24}$$

where α_r and β_r are fitting parameters, which model the dominant effects between carrier localization and carrier tunneling corresponding to the amount of roughness amplitude. Equation (6.24) models the increase in BTBT due to the edge states in the bandgap by summing and normalizing the neighboring GNRs. The integer m has the value of the ratio $\Delta W/\sqrt{3}a_{\text{cc}}$. ΔW_c = 0.04 is the critical width fluctuation amplitude. The BTBT through edge states in the bandgap leads to the increase in the net transport of carriers from source to drain; consequently, the increase in the conduction of GNRFETs for $\Delta W_c > \Delta W$. By increasing the roughness amplitude larger than ΔW_c, however, the tunneling of carriers occurs mostly between the localized states without a net transport of carriers from the source to the drain regions. The exponential decrease in device conductivity in the localization regime has been modeled with both the increase in the sub-band energies in Eq. (6.23) and in the exponential term in Eq. (6.24).

6.3.5 Extracting Fitting Parameters

The fitting parameters in our developed GNRFET model have been extracted by matching its transfer characteristics with numerical data in regards to the following procedure:

1. Obtain the numerical data from NEGF simulation for bias conditions and the device geometries related to the BTBT phenomena.
2. Obtain the analytical results using the developed model for the same bias conditions and device geometries for a given fitting parameter.
3. Change a fitting parameter according to its broad range to determine the best value in which the root mean square error (RMSE) between the numerical and analytical data is minimized.

The above procedure is relatively quick since numerical data, which can be computationally intensive, need to be calculated only once for a device setting. Then the analytical model provides prompt results that need to be repeated to search for the best value of the fitting parameters. In addition, all the fitting parameters in this work are used for modeling BTBT from drain to channel with only dominant effects on the total current of the FET structure for narrow-bandgap GNRs and under bias conditions of high V_{DS} and low V_{GS} (see Fig. 6.3b). As such, the fitting procedure can be limited to those bias conditions (subthreshold region) and wider GNR channels, which corresponds to higher RMSE between numerical and analytical results.

The dependence of fitting parameters on the GNR width can be eliminated by including another fitting dimension for semiconducting GNR chiralities such as GNR(N,0) shown in Fig. 6.5. The figure shows an example of a 2D fitting procedure, in which all the analytical results versus gate voltage and GNR widths are repeated for different values of a fitting parameter, e.g., δ, approaching the best fit with the smallest average RMSE in two dimensions.

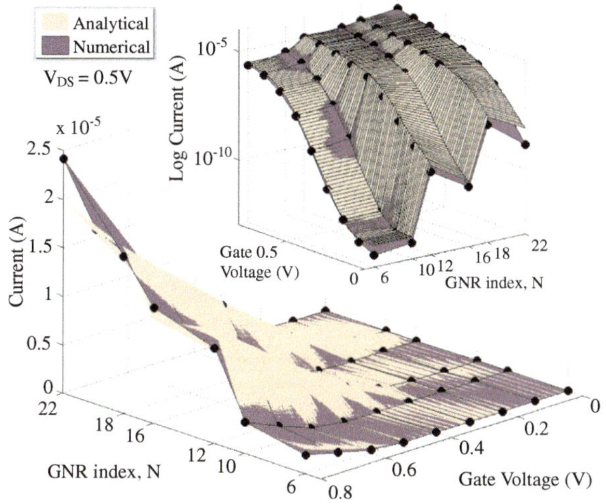

Figure 6.5 An example of 2D fitting of continuous analytical results to the variety of discrete numerical data from the NEGF simulation, which has been used to obtain the fitting parameter δ = 0.05 [61].

The fitting procedure is repeated until the RMSE reaches ±10% and ±20% of the numerical data for fitting parameters related to ideal GNRs and non-ideal GNRs, respectively. The maximum error values are our assigned acceptable errors for obtaining the general fitting parameters independent of device dimensions, which correspond to the GNRFETs with 16 nm channel length of GNR(24,0) at V_{DS} = 0.8 V and V_{GS} = 0.05 (significant BTBT phenomena). Two fitting parameters, l_{relax} and δ in Eq. (6.15), are associated with ballistic transport in ideal GNRs while γ, α_r, and β_r in Eqs. (6.23) and (6.24) are related to the non-ballistic transport modeling of carriers in rough-edge GNRs. Table 6.1 shows the searching ranges and obtained values of the fitting parameters, as well as the maximum errors with regard to the numerical data. For example, by altering l_{relax} in a range from 10 nm to 100 nm, the analytical results match the numerical data in the subthreshold region within the acceptable RMSE at l_{relax} = 40 nm. A similar method was used to obtain the l_{relax} of the Stanford CNTFET model [24]. The fitting parameters in our GNRFET model only need to be obtained just once and their values are valid for all the device geometries and LER in this study, as indicated in Table 6.2.

Table 6.1 Obtained values of fitting parameters, the searching ranges and step, along with the corresponding errors

Fitting Parameter	Obtained Value	Searching Range	Searching Step	% RMSE
l_{relax}	40 nm	10–100 nm	5 nm	<10%
δ	0.05	0.01–0.1	0.025	
γ	2.9	0–3	0.5, then 0.1	
α_r	6×10^{-6}	From 0	Different (Nonlinear)	<20%
β_r	0.0145	From 0	0.01, 0.001 and 0.0001	

Table 6.2 Scope of the device dimension and line-edge roughness in GNRFET model

Device Dimension	**GNR Channel Length (L_{CH})**	10–45 nm
	GNR Channel Width (W_{GNR}, GNR(N, 0))	~0.87 nm (N = 6) to ~3.18 nm (N = 25)
	Oxide Thickness (t_{ox})	0.5–2.0 nm
Line-edge Roughness	**Roughness Amplitude (ΔW)**	0.0–0.2
	Roughness Correlation (ΔL)	3–10 nm

6.4 Model Validation

6.4.1 Comparing with Computational NEGF Formalism

The dimension and bias conditions of most experimental works are much larger than the target ranges of GNRFETs for emerging technology. For example, the GNR width (W_{GNR}) and gate voltage, (V_{G}) are mostly studied in the range of 10–150 nm and 10–50 V, respectively [48–50]. Therefore, we have validated the accuracy of our developed analytical model against the device-level atomistic numerical simulation based on the NEGF formalism as described for ideal GNRFETs in Ref. [22] and for GNRFET with LER in Ref. [46]. Figures 6.6a and 6.6b show the I_{DS}–V_{GS} characteristic of GNRFETs with three different GNR indices of N = 6, 12, and 18 for drain voltages of 0.1 V and 0.5 V, respectively. It is shown that our analytical model agrees well with numerical simulations. The device dimensions in the simulation are L_G = 16 nm, W_G = 2 nm, and t_{ox} = 1 nm. For comparison with device-level simulation, the Fermi level due to the work function difference between metallic gate and graphene is set to E_F = 0, while its nominal value is kept equal to E_F = 0.4 eV for the other simulations. It can be observed that increasing the GNR width or drain voltage leads to the increase in the OFF-current as both can provide more sub-bands to incorporate in the BTBT of carriers from drain to channel. Figure 6.6c shows the $I_{DS}-V_{DS}$ characteristics of GNRFETs at different gate voltages, demonstrating that the results of the analytical model in this work are in correlation with those of numerical simulations.

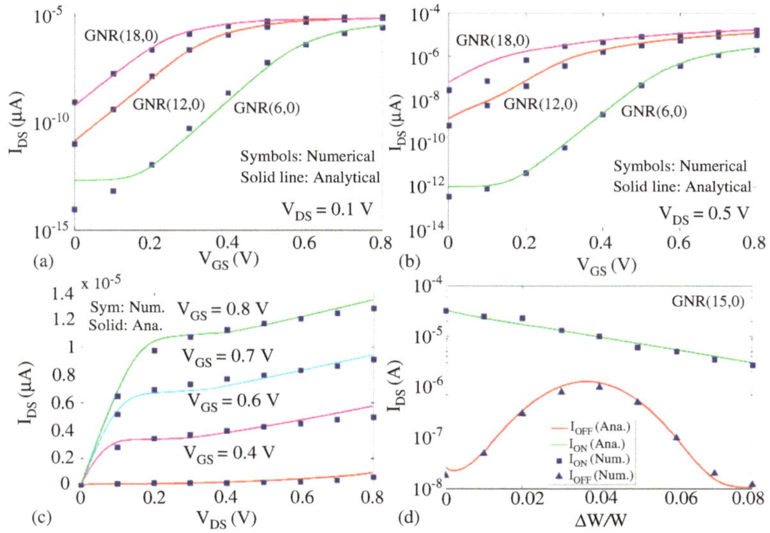

Figure 6.6 I_{DS}–V_{GS} characteristic of GNRFETs with three different GNR indices of N = 6, 12, and 18 for (a) V_{DS} = 0.1 V and (b) V_{DS} = 0.5 V, respectively; (c) I_{DS}–V_{DS} characteristics of GNRFETs at different gate voltages. *Note*: The numerical results are shown with dotted symbols and the analytical results with solid lines. (d) I_D (ON) and I_D (OFF) versus $\Delta W/W$ for L_{CH} = 16 nm [61].

Figure 6.6d shows the effect of LER on the OFF-state and ON-state currents of GNRFETs with the channel of GNR(15,0). It can be observed that the analytical model agrees very well with the numerical simulations, which can be obtained at the expense of long computational time by statistical averaging many GNR samples with the same roughness parameters [14]. Furthermore, the GNRFETs with perfectly smooth edges can have more than three orders of magnitude difference between ON- and OFF-currents. The OFF-current increases by increasing the roughness amplitude, as the formation of some localized states in the bandgap significantly increases the BTBT of carriers. Conversely, the ON-current decreases by increasing the roughness amplitude due to the increase in carrier scattering associated with LER. By increasing the LER, the carrier transport becomes more diffusive and the trend continues up to the critical roughness amplitude (ΔW_c). For LER amplitude larger than ΔW_c, the BTBT of carriers through the localized edge states is dominated by the localization of carriers in those states, while the backscattering of carriers occurs due to the localized potential barriers of edge disorders. The critical roughness amplitude corresponds with the maximum OFF-current and the minimum value of the $I_{\mathrm{ON}}/I_{\mathrm{OFF}}$ ratio. By increasing the roughness amplitude, the OFF-current is decreased as the carriers transport takes place by tunneling between the localized edge states, leading to the reduction in the effective net transport of carriers from drain to channel. It is worth mentioning that the uniform edge roughness has been assumed in some works [51, 52] by defining the roughness percentage p, which neglects the correlation between edge disorders. This assumption underestimates the effect of LER; thereby, only the initial increase in OFF-current due to the BTBT phenomena has been exhibited, while larger LER reduces the OFF-current similar to the ON-state current.

6.4.2 Comparing with Many-Body Problem

It has been shown both experimentally [13, 53] and theoretically [14] that strong localization can appear in a single-layer GNR for large LER, known as Anderson-type localization [54]. Edge disorder increases the number of localized carriers and their corresponding DOS at the edges, which blocks the conductive paths of carrier transport through the graphene nanoribbon. This can potentially increase the importance of electron–electron interaction for carrier transport, such that the validity of single-particle calculations for this problem needs to be justified. The effects of electron–electron interactions on carrier transport need to be investigated using numerical simulations; then the corresponding results can be incorporated in the analytical calculations. Many samples need to be created with given roughness parameters, and each must be simulated by a self-consistent iteration algorithm between the carrier transport problem using the NEGF formalism and the electrostatic problem using the Poisson equation. Incorporating the electron–electron interaction in the problem will add another iteration algorithm between the Green's function and the electron–electron self-energy [55]. From the obtained numerical data and using the equation $<T(E)> = M(E)/(1+L_{CH}/\lambda(E))$, where $M(E)$ is the number of active conduction channels and $<T(E)>$ is the average transmission probability, the MFP of the carriers can be extracted to investigate the role of LER scattering and the electron–electron interaction in the transport problem [55].

Figure 6.7 shows the MFPs associated with the LER and the electron–electron interaction. In the simulation, the edge roughness is set to be highly correlated, ΔL = 3 nm, which reduces the MFP below the nanometer range and the electron–electron scattering rate is assumed 50 ps for a momentum conserving interaction [56]. It can be seen that the analytical results of Eq. (6.20) for LER are in agreement with the numerical simulations and are in correspondence with Fermi's golden rule ($\lambda^{LER} \propto 1/\Delta W^2$) [45]. By incorporating the electron–electron interaction, there is a deviation in the total MFP of carriers at small roughness amplitudes. The total MFP is dominated by the effects of the electron–electron interaction at small roughness, while the LER scattering is the dominant scattering mechanism at larger roughness amplitudes. Electron–electron interactions are known to have significant effects on the carrier transport in materials with 1D confinement, e.g., GNR interconnects. As the GNRs in this application need to be long, the 1D problem is very strong and carrier transport is deep in the localization regime [57]. The carrier transport can only take place by tunneling between the localized edge states [58], which results in a significant reduction in transmission probability [54]. However, the GNR length as a channel material is 16 nm in our simulation, which is a typical channel length for emerging transistors and the GNR width is ~1.6 nm; thus, the strength of the 1D problem corresponding to the ratio of the GNR channel length to GNR width ($W/L \approx 1/10$) is not large enough to make electron–electron interaction a considerable portion for most roughness amplitudes. However, the soundness of single-particle calculations with regard to the width-to-length ratio of GNR is still controversial in the literature, such that charge localization due to strong edge roughness has been reported in some works [59, 60] as a responsible mechanism for Coulomb blockade and the formation of bottlenecks for carrier transport in GNR channels. Further research is required to understand the exact mechanism for charge localization along the edge. Particularly, the way the edge disorder changes the potential landscape in a graphene nanoribbon so that electrons prefer to be located along the edge rather than transported in the bulk needs to be investigated.

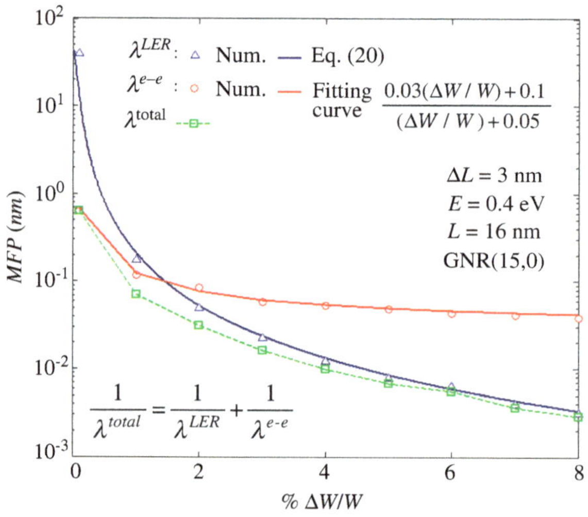

Figure 6.7 Mean free path (MFP) of line-edge roughness and electron–electron interaction. *Note*: Symbol shows the numerical results, and solid curves show the analytical results from Eq. (6.20) for the MFP of line-edge roughness together with the fitting equation for the extracted MFP of the electron–electron interaction [61].

6.5 Effect of Edge Roughness on Device Characteristic

6.5.1 Transfer Characteristics of GNRFETs

Figure 6.8 shows the I_{DS}–V_{GS} characteristic of three GNRFETs with the channels of GNR(9,0), GNR(15,0), and GNR(24,0) at V_{DS} = 0.25 V for various LER amplitudes and constant roughness correlation of ΔL = 10 nm. The narrower ribbon, i.e., GNR(9,0), has lower OFF-current than GNR(15,0) and GNR(24,0) for all LER amplitudes. At V_{DS} = 0.25 V, the BTBT of the GNRFET with the larger bandgap, i.e., GNR(9,0), is very small; thus, the edge states introduced in the bandgap by increasing LER amplitude cannot lead to a significant increase in the OFF-state current (V_{GS} = 0). This increasing LER amplitude only increases the scattering of carriers, consequently reducing the currents. For GNR(24,0), however, the bandgap is small, and the BTBT can be significant at V_{DS} = 0.25 V. Thus, the OFF-current initially increases by increasing the roughness amplitude up to the critical roughness amplitude (ΔW_c = 0.04), then

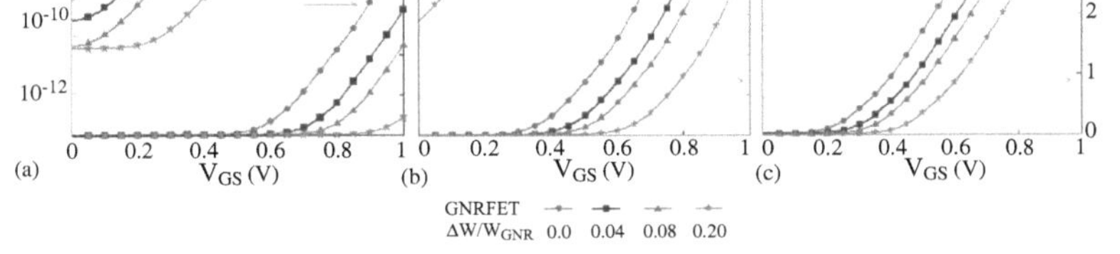

Figure 6.8 I_{DS}–V_{GS} characteristic of three GNRFETs with the channel of (a) GNR(9,0); (b) GNR(15,0); and (c) GNR(24,0) at V_{DS} = 0.25 V. The ideal edge GNRFETs and non-ideal GNRFETs with edge roughness variations of 0.04, 0.08, and 0.2 are shown [61].

decreases at larger roughness amplitude due to the backscattering of carriers from the localized edge states. However, the edge roughness continuously increases the scattering of thermionic carriers at the ON-state; thereby, the reduction in the ON-current is continuous, correspondingly. Both increasing the OFF-current and decreasing the ON-current can deteriorate the device performance at a critical roughness amplitude. For large LER amplitude, however, the main source of performance degradation is the significant decrease in the ON-current of GNRFETs. The same effect of roughness amplitude can be observed in narrower GNRs (larger bandgaps) and at higher V_{DS}.

6.5.2 OFF-State Characteristics of GNRFETs

Figure 6.9 shows the OFF-current, ON-current, and I_{ON}/I_{OFF} ratio as a function of the GNR width (semiconducting GNR index $N = 3p$, where p is an integer). Both the ON- and OFF-currents of ideal GNRFETs increase by widening the GNR width, as larger numbers of sub-bands can contribute to carrier transport by decreasing the bandgap at a given supply voltage. The I_{ON}/I_{OFF} ratio is increased by decreasing the GNR width, as the leakage current associated with the BTBT phenomena is significantly decreased by increasing the bandgap. At the scaled supply voltage (V_{DD} = 0.5 V), GNR(6,0) with a larger bandgap is not fully at the ON-state, resulting in smaller I_{ON} /I_{OFF} ratio. GNRFETs demonstrate larger OFF-current at the critical LER amplitude of ($\Delta W/W_{GNR}$ = 0.04 due to the maximum number of BTBT from drain to channel occurring through the generated edge states in the bandgap. While I_{OFF} is larger and I_{ON} is smaller than the ideal narrow GNRFETs due to the decrease in carrier scattering at the ON-state, which leads to significant deterioration of the I_{ON}/I_{OFF} ratio for ($\Delta W/W_{GNR}$ = 0.04. Larger LER amplitudes, e.g., ($\Delta W/W_{GNR}$ = 0.1, can set the carrier transport in the localization regime such that both the OFF- and ON-currents are reduced resulting from the decrease in the MFP of edge scattering and the increase in the backscattering of carriers occurred by a larger number of localized edge states.

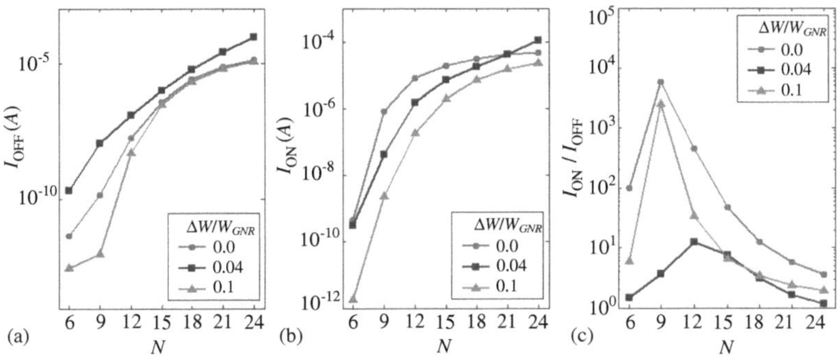

Figure 6.9 (a) OFF-current, (b) ON-current, and (c) I_{ON}/I_{OFF} ratio as a function of armchair GNR index *N*. *Note*: V_{DD} = 0.5 V [61].

6.6 Summary

We present a physics-based analytical model compatible with SPICE for the circuit simulation of GNRFETs. The carrier charge and current have been analytically calculated and compared with the numerical data obtained by the NEGF formalism. We presented the device-level performance of non-ideal GNRFETs in edge-enhanced BTBT and localization regimes for various roughness amplitudes. Corresponding to these two regimes, I_{OFF} is initially increased and then decreased, while I_{ON} is continuously decreased by increasing the roughness amplitude. The BTBT and the corresponding increase in the OFF-current can be reduced by accurately choosing the supply voltage for a given GNR width. The LER increases the threshold voltage and limits the advantage of bandgap engineering of GNRFETs for scaling down the supply voltages. The LER in GNRFETs associated with the lack of atomic-scale precision in current patterning technique, which significantly degrades the performance of GNRFETs.

References

1. Wilson, L. (2013). *Semiconductor Industry Association*, International Technology Roadmap for Semiconductors (ITRS).

2. Novoselov, K. S., Fal, V., Colombo, L., Gellert, P., Schwab, M., and Kim, K. (2012). A roadmap for graphene, *Nature*, **490**, pp. 192–200.

3. Obeng, Y., De Gendt, S., Srinivasan, P., Misra, D., Iwai, H., Karim, Z., Hess, D., and Grebel, H. (2009) Graphene and emerging materials for post-CMOS applications, *ECS Transactions*, **19**.

4. Novoselov, K., Geim, A. K., Morozov, S., Jiang, D., Katsnelson, M., Grigorieva, I., Dubonos, S., and Firsov, A. (2005). Two-dimensional gas of massless Dirac fermions in graphene, *Nature*, **438**, pp. 197–200.

5. Cooper, D. R., D'Anjou, B., Ghattamaneni, N., Harack, B., Hilke, M., Horth, A., Majlis, N., Massicotte, M., Vandsburger, L., and Whiteway, E. (2012). Experimental review of graphene, *International Scholarly Research Notices*, **2012**, p. 501686.

6. Schwierz, F. (2013). Graphene transistors: Status, prospects, and problems, *Proceedings of the IEEE*, **101**, pp. 1567–1584.

7. Harada, N., Sato, S., and Yokoyama, N. (2013). Theoretical investigation of graphene nanoribbon field-effect transistors designed for digital applications, *Japanese Journal of Applied Physics*, **52**, p. 094301.

8. Johari, Z., Hamid, F., Tan, M. L. P., Ahmadi, M. T., Harun, F., and Ismail, R. (2013). Graphene nanoribbon field effect transistor logic gates performance projection, *Journal of Computational and Theoretical Nanoscience*, **10**, pp. 1164–1170.

9. Poljak, M., Wang, K. L., and Suligoj, T. (2015). Variability of bandgap and carrier mobility caused by edge defects in ultra-narrow graphene nanoribbons, *Solid-State Electronics*, **108**, pp. 67–74.

10. Michetti, P. and Iannaccone, G. (2010). Analytical model of one-dimensional carbon-based Schottky-barrier transistors, *IEEE Transactions on Electron Devices*, **57**, pp. 1616–1625.

11. Choudhury, M., Yoon, Y., Guo, J., and Mohanram, K. (2008). Technology exploration for graphene nanoribbon FETs, *2008 45th ACM/IEEE Design Automation Conference*, Anaheim, CA, pp. 272–277.

12. Chen, Y.-Y., Rogachev, A., Sangai, A., Iannaccone, G., Fiori, G., and Chen, D. (2013). A SPICE-compatible model of graphene nano-ribbon field-effect transistors enabling circuit-level delay and power analysis under process variation, 2013 *Design, Automation & Test in Europe Conference & Exhibition (DATE)*, Grenoble, France, pp. 1789–1794.

13. Xu, G., Torres Jr, C. M., Tang, J., Bai, J., Song, E. B., Huang, Y., Duan, X., Zhang, Y., and Wang, K. L. (2011). Edge effect on resistance scaling rules in graphene nanostructures, *Nano Letters*, **11**, pp. 1082–1086.

14. Yazdanpanah, A., Pourfath, M., Fathipour, M., Kosina, H., and Selberherr, S. (2012). A numerical study of line-edge roughness scattering in graphene nanoribbons, *IEEE Transactions on Electron Devices*, **59**, pp. 433–440.

15. Yoon, Y., Fiori, G., Hong, S., Iannaccone, G., and Guo, J. (2008). Performance comparison of graphene nanoribbon FETs with Schottky contacts and doped reservoirs, *IEEE Transactions on Electron Devices*, **55**, pp. 2314–2323.

16. Kang, J., Sarkar, D., Khatami, Y., and Banerjee, K. (2013). Proposal for all-graphene monolithic logic circuits, *Applied Physics Letters*, **103**, p. 083113.

17. Owlia, H. and Keshavarzi, P. (2014). Investigation of the novel attributes of a double-gate graphene nanoribbon FET with AlN high-κ dielectrics, *Superlattices and Microstructures*, **75**, pp. 613–620.

18. Oh, J. G., Hong, S. K., Kim, C.-K., Bong, J. H., Shin, J., Choi, S.-Y., and Cho, B. J. (2014). High performance graphene field effect transistors on an aluminum nitride substrate with high surface phonon energy, *Applied Physics Letters*, **104**, p. 193112.

19. Konar, A., Fang, T., and Jena, D. (2009). Effect of high-κ dielectrics on charge transport in graphene, *arXiv preprint arXiv:0902.0819*.

20. Banadaki, Y. and Srivastava, A. (2015). Scaling effects on static metrics and switching attributes of graphene nanoribbon FET for emerging technology, *IEEE Transactions on Emerging Topics in Computing*, **3**, pp. 458–469.

21. Grassi, R., Poli, S., Gnani, E., Gnudi, A., Reggiani, S., and Baccarani, G. (2009). Tight-binding and effective mass modeling of armchair graphene nanoribbon FETs, *Solid-State Electronics*, **53**, pp. 462–467.

22. Banadaki, Y. M. and Srivastava, A. (2015). Investigation of the width-dependent static characteristics of graphene nanoribbon field effect transistors using non-parabolic quantum-based model, *Solid-State Electronics*, **111**, pp. 80–90.

23. Kliros, G. S. (2012). Scaling effects on the gate capacitance of graphene nanoribbon transistors, *CAS 2012 (International Semiconductor Conference)*, Sinaia, pp. 83–86.

24. Deng, J. and Wong, H. P. (2007). A compact SPICE model for carbon-nanotube field-effect transistors including nonidealities and its application—Part I: Model of the intrinsic channel region, *IEEE Transactions on Electron Devices*, **54**, pp. 3186–3194.

25. Gholipour, M., Chen, Y.-Y., Sangai, A., and Chen, D. (2014). Highly accurate SPICE-compatible modeling for single-and double-gate GNRFETs with studies on technology scaling, *2014 Design, Automation & Test in Europe Conference & Exhibition (DATE)*, Dresden, pp. 1–6.
26. Calculator, M.C. http://www.technick.net/public/code/cp_dpage.php?aiocp_dp=util_pcb_imp_microstrip_embed.
27. Garg, R., Bahl, I., and Bozzi, M. (2013). *Microstrip Lines and Slotlines*, Artech House.
28. Wang, J. and Lundstrom, M. (2002). Does source-to-drain tunneling limit the ultimate scaling of MOSFETs? *Digest. International Electron Devices Meeting*, San Francisco, CA, USA, pp. 707–710.
29. Datta, S. (2005). *Quantum Transport: Atom to Transistor*, Cambridge University Press.
30. Kane, E. O. (1961). Theory of tunneling, *Journal of Applied Physics*, **32**, pp. 83–91.
31. Geist, J. and Lowney, J. (1981). Effect of band-gap narrowing on the built-in electric field in n-type silicon, *Journal of Applied Physics*, **52**, pp. 1121–1123.
32. Geim, A. K. and Novoselov, K. S. (2007). The rise of graphene, *Nature Materials*, **6**, pp. 183–191.
33. Akturk, A. and Goldsman, N. (2008). Electron transport and full-band electron-phonon interactions in graphene, *Journal of Applied Physics*, **103**, p. 053702.
34. Chen, J.-H., Jang, C., Xiao, S., Ishigami, M., and Fuhrer, M. S. (2008). Intrinsic and extrinsic performance limits of graphene devices on SiO_2, *Nature Nanotechnology*, **3**, pp. 206–209.
35. Yang, Y. and Murali, R. (2010). Impact of size effect on graphene nanoribbon transport, *IEEE Electron Device Letters*, **31**, pp. 237–239.
36. Zhao, P., Choudhury, M., Mohanram, K., and Guo, J. (2008). Computational model of edge effects in graphene nanoribbon transistors, *Nano Research*, **1**, pp. 395–402.
37. Lundstrom, M. and Guo, J. (2006). *Nanoscale Transistors: Device Physics, Modeling and Simulation*, Springer US.
38. Perebeinos, V., Tersoff, J., and Avouris, P. (2005). Electron-phonon interaction and transport in semiconducting carbon nanotubes, *Physical Review Letters*, **94**, p. 086802.
39. Rakheja, S., Kumar, V., and Naeemi, A. (2013). Evaluation of the potential performance of graphene nanoribbons as on-chip interconnects, *Proceedings of the IEEE*, **101**, pp. 1740–1765.

40. Wang, X., Ouyang, Y., Li, X., Wang, H., Guo, J., and Dai, H. (2008). Room-temperature all-semiconducting sub-10-nm graphene nanoribbon field-effect transistors, *Physical Review Letters*, **100**, p. 206803.

41. Areshkin, D. A., Gunlycke, D., and White, C. T. (2007). Ballistic transport in graphene nanostrips in the presence of disorder: Importance of edge effects, *Nano Letters*, **7**, pp. 204–210.

42. Gunlycke, D., Areshkin, D., and White, C. (2007). Semiconducting graphene nanostrips with edge disorder, *Applied Physics Letters*, **90**, p. 142104.

43. Gunlycke, D. and White, C. (2010). Scaling of the localization length in armchair-edge graphene nanoribbons, *Physical Review B*, **81**, p. 075434.

44. Goodnick, S., Ferry, D., Wilmsen, C., Liliental, Z., Fathy, D., and Krivanek, O. (1985). Surface roughness at the Si(100)–SiO_2 interface, *Physical Review B*, **32**, p. 8171.

45. Fang, T., Konar, A., Xing, H., and Jena, D. (2008). Mobility in semiconducting graphene nanoribbons: Phonon, impurity, and edge roughness scattering, *Physical Review B*, **78**, p. 205403.

46. Goharrizi, A. Y., Pourfath, M., Fathipour, M., Kosina, H., and Selberherr, S. (2011). An analytical model for line-edge roughness limited mobility of graphene nanoribbons, *IEEE Transactions on Electron Devices*, **58**, pp. 3725–3735.

47. Luisier, M. and Klimeck, G. (2009). Performance analysis of statistical samples of graphene nanoribbon tunneling transistors with line edge roughness, *Applied Physics Letters*, **94**, p. 223505.

48. Bai, J., Duan, X., and Huang, Y. (2009). Rational fabrication of graphene nanoribbons using a nanowire etch mask, *Nano Letters*, **9**, pp. 2083–2087.

49. Lin, M.-W., Ling, C., Zhang, Y., Yoon, H. J., Cheng, M. M.-C., Agapito, L. A., Kioussis, N., Widjaja, N., and Zhou, Z. (2011). Room-temperature high on/off ratio in suspended graphene nanoribbon field-effect transistors, *Nanotechnology*, **22**, p. 265201.

50. Huang, C.-H., Su, C.-Y., Okada, T., Li, L.-J., Ho, K.-I., Li, P.-W., Chen, I.-H., Chou, C., Lai, C.-S., and Samukawa, S. (2013). Ultra-low-edge-defect graphene nanoribbons patterned by neutral beam, *Carbon*, **61**, pp. 229–235.

51. Yoon, Y. and Guo, J. (2007). Effect of edge roughness in graphene nanoribbon transistors, *Applied Physics Letters*, **91**, p. 073103.

52. Fiori, G. and Iannaccone, G. (2007). Simulation of graphene nanoribbon field-effect transistors, *IEEE Electron Device Letters*, **28**, pp. 760–762.

53. Sprinkle, M., Ruan, M., Hu, Y., Hankinson, J., Rubio-Roy, M., Zhang, B., Wu, X., Berger, C., and De Heer, W. A. (2010). Scalable templated growth of graphene nanoribbons on SiC, *Nature Nanotechnology*, **5**, pp. 727–731.

54. Evaldsson, M., Zozoulenko, I. V., Xu, H., and Heinzel, T. (2008). Edge-disorder-induced Anderson localization and conduction gap in graphene nanoribbons, *Physical Review B*, **78**, p. 161407.

55. Kahnoj, S. S., Touski, S. B., and Pourfath, M. (2014). The effect of electron-electron interaction induced dephasing on electronic transport in graphene nanoribbons, *Applied Physics Letters*, **105**, p. 103502.

56. Li, X., Barry, E., Zavada, J., Nardelli, M. B., and Kim, K. (2010). Influence of electron-electron scattering on transport characteristics in monolayer graphene, *Applied Physics Letters*, **97**, p. 082101.

57. Lherbier, A., Biel, B., Niquet, Y.-M., and Roche, S. (2008). Transport length scales in disordered graphene-based materials: Strong localization regimes and dimensionality effects, *Physical Review Letters*, **100**, p. 036803.

58. Datta, S. (1997). *Electronic Transport in Mesoscopic Systems*, Cambridge University Press.

59. Bischoff, D., Varlet, A., Simonet, P., Eich, M., Overweg, H., Ihn, T., and Ensslin, K. (2015). Localized charge carriers in graphene nanodevices, *Applied Physics Reviews*, **2**, p. 031301.

60. Sols, F., Guinea, F., and Neto, A. C. (2007). Coulomb blockade in graphene nanoribbons, *Physical Review Letters*, **99**, p. 166803.

61. Banadaki, Y. M. and Srivastava, A. (2016). Effect of edge roughness on static characteristics of graphene nanoribbon field effect transistor, *Electronics*, **5**, p. 11.

Chapter 7

Graphene-Based Circuit Design

7.1 Introduction

In this chapter, we investigate the effect of edge roughness on the power and delay performance of graphene nanoribbon field-effect transistors (GNRFETs) by developing an analytical model of multi-GNR channel FETs in SPICE. We describe the advantage and challenge of all-graphene circuits and investigate the delay, power, and energy-delay product (EDP) of GNRFETs for various edge roughness amplitudes and supply voltages. We compare GNRFET performance with 16 nm node silicon complementary metal–oxide–semiconductor (Si-CMOS) by designing a 5-stage fanout-of-4 buffer chain.

The chapter continues by designing a new energy-efficient exclusive-NOR (XNOR) and exclusive-OR (XOR) gate based on clocked adiabatic logic (CAL), which reduces the power density by nearly 65% in comparison with the earlier CAL design in 45 nm technology node. Then we substitute conventional CMOS in this design with GNRFETs for a set of SPICE simulation. We show that GNRFETs allow scaling of the supply voltage, which results in nearly three times reduction in power density. The on-chip power density of new XNOR/XOR gates remains below the limit reported by the International Technology Roadmap for Semiconductors (ITRS) up to the operation frequency of about 1 GHz at the scaled supply voltage equal to 0.7 V.

Graphene Nanostructures: Modeling, Simulation, and Applications in Electronics and Photonics
Yaser M. Banadaki and Safura Sharifi

ISBN 978-981-4800-36-5 (Hardcover), 978-0-429-02221-0 (eBook)
www.jennystanford.com

7.2 All-Graphene Circuits

Figure 7.1a shows the three-dimensional (3D) view of an all-graphene circuit for an example of inverter chains together with its circuit implementation [1]. In this structure, the devices and interconnects are concurrently fabricated by monolithically patterning a single sheet of graphene. This would bring some relief from the contact resistance of metal-to-graphene contacts as the material for producing both devices and interconnects is graphene [2, 3]. Pattering GNR with larger width and along zigzag-edge orientation results in GNR interconnects with very small bandgap (Fig. 7.1b) [4]. For using graphene as a channel material and local interconnects, both the width and the edge type of GNRs are critically important. This can potentially reduce the complex fabrication process for local interconnect in nanoscale dimensions, leading to ultra-dense and thin integrated circuits [5]. While a modern-day CMOS circuit has approximately 10 interconnect layers, this structure can reduce the number of intra-layer local interconnects for gate-level design. However, it is not possible to get rid of metal contacts and interconnects completely since the gate and source/drain electrodes cannot share the same graphene sheet. The contact resistances of metal–graphene junctions are reported in the range of 1 kΩ to 100 kΩ, with a nominal value of R_{via} = 20 kΩ for via connections with 50 nm width [6].

GNR is atomically thin in the vertical direction and quantum-mechanically confined in the transverse direction, leading to a small drive current of GNRFETs. Fabricating multiple parallel armchair GNRs in an array connected to the same wide zigzag GNRs can increase the drive current, leading to better switching attribute for high-performance applications [7]. As such, the number of GNR channels (N_r) in a GNRFET can correspond to an integer increment of W/L in conventional CMOS. As the bandgap of GNRs can be inversely changed by the GNR width (W_{GNR}), it provides another degree of freedom for designers to use bandgap engineering in GNRFET circuits [8]. A multi-channel GNRFET can be interpreted as several parallel individual GNRFETs, as shown in Fig. 7.1c. The length of local GNR interconnects between transistors within logic gates is much shorter than the mean free path of graphene. In addition, the GNR interconnects are wide enough to maintain the condition of very small bandgap. Therefore, the effects of edge roughness on interconnects can be neglected. Also the resistance and capacitance of local GNR interconnects can be assumed negligible in the first-order models [9].

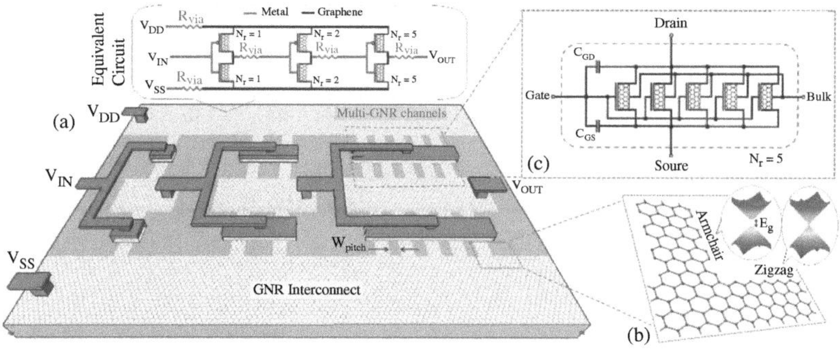

Figure 7.1 (a) 3D view of an all-graphene circuit for an example of inverter chains together with its circuit implementation. The contact resistances of via connections are also shown corresponding to the layout design. (b) Narrow armchair-edge GNRs and wide zigzag-edge GNRs are used as channel material and local interconnects, respectively. (c) Sub-circuit of multi-channel GNRFET with five channels connected as five parallel individual GNRFETs [1].

7.3 Graphene Inverter

A complementary logic inverter can be designed as one of the main building blocks of integrated circuits by integrating two complementary GNRFETs. The voltage transfer characteristics (VTCs) and waveform of a GNRFET-based inverter are evaluated for various line-edge roughness (LER) amplitudes and supply voltages, as shown in Fig. 7.2. The large-edge roughness ($\Delta W/W_{\text{GNR}}$ = 0.2) significantly deteriorates the VTC of GNRFETs at V_{DD} = 0.7 V. The bandgap of the GNR(15,0) channel is relatively small ($E_g \cong 0.25$ eV), and thus the band-to-band tunneling (BTBT) between the channel and drain regions can be high at large supply voltages. This can significantly increase the effect of edge states in the bandgap, resulting in orders of magnitudes higher leakage currents. However, scaling down the supply voltage to 0.3 V reduces the contribution of edge states on the BTBT phenomena and makes the curve almost unaffected with increasing LER amplitude.

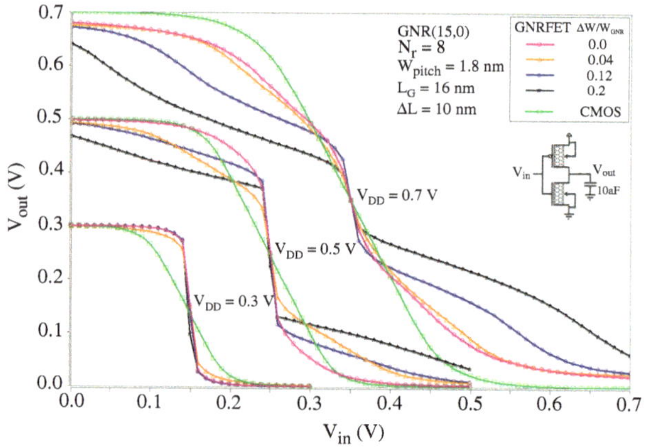

Figure 7.2 Voltage transfer characteristics (VTCs) of GNRFET-based inverters for various line-edge roughness amplitudes and supply voltages. For comparison, the VTCs of Si-CMOS inverters with the same physical size and parameters are also shown [1].

The noise margin (NM), the maximum voltage gain of inverter (A_{INV}), and the output voltage swing (V_{OS}) are three functional criteria of an inverter that evaluate its reliability and robustness to a superimposed noise on a digital signal without causing a malfunction of an inversion operation. Figure 7.3a shows the NM of GNRFETs versus roughness amplitudes for three supply voltages of V_{DD} = 0.3 V, 0.5 V, and 0.7 V. By scaling the supply voltage down to V_{DD} = 0.3 V, the NM of GNRFETs improves above that of Si-CMOS at a nominal supply voltage of V_{DD} = 0.7 V for 16 nm technology node. At high supply voltage, the NM of a GNRFET inverter is significantly decreased by the LER amplitude, such that the NM regions are totally diminished at $\Delta W/W_{GNR}$ = 0.16. Decreasing the length of edge roughness correlation from ΔL = 10 nm to 3 nm increases the dependence of the NM on the LER amplitude. This shifts the zero NM point to $\Delta W/W_{GNR}$ = 0.1 because of larger carrier scattering, followed by carrier localization. The maximum voltage gain of inverters (A_v) can be defined by the maximum slope of the VTC in the transition region. It has been shown in Fig. 7.3b that the gain of a GNRFET-based inverter is increased to 30 by scaling down the supply voltage to V_{DD} = 0.3 V, which shows above 10 times larger value than that of Si-CMOS and almost independent of edge roughness amplitude.

It can be observed in Fig. 7.3c that the output voltage swing (V_{OS}) is also high at the scaled supply voltage, reaching that of Si-CMOS values close to 100%. At small supply voltages, all three curves of NM, A_{INV}, and V_{OS} are slightly changed by the roughness amplitude ratio ($\Delta W/W_{GNR}$). The small deteriorations in the VTC attributes around the critical roughness amplitude (ΔW_c) are due to the edge-enhanced BTBT and the corresponding increase in the leakage current.

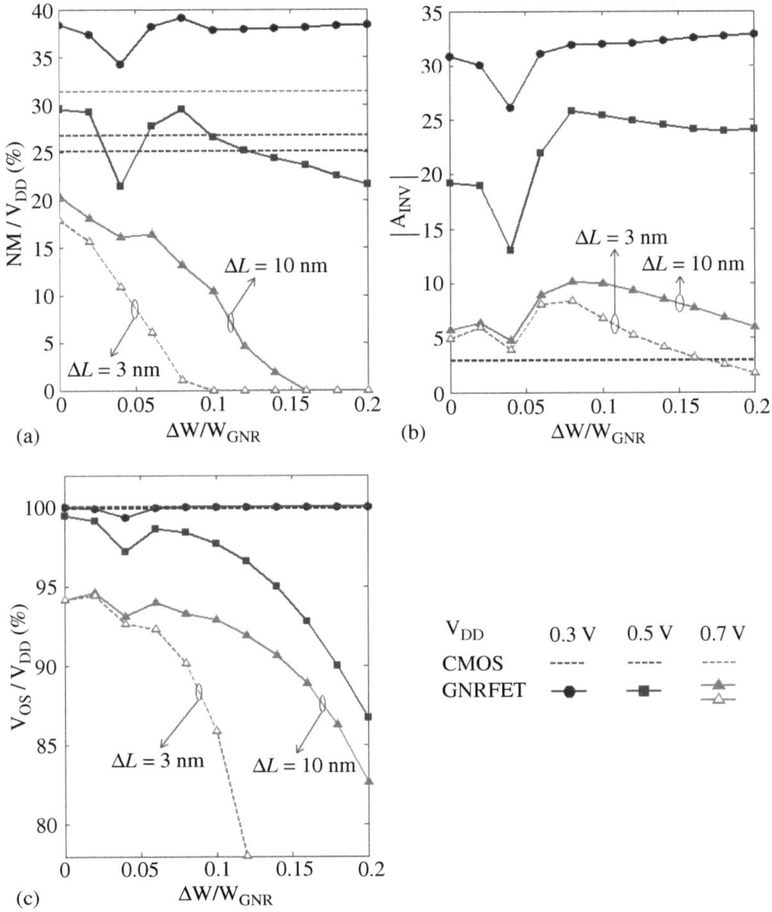

Figure 7.3 (a) Noise margin (NM), (b) maximum voltage gain of inverters (A_{INV}), and (c) output voltage swing (V_{OS}) of GNRFETs versus roughness amplitudes for three supply voltages of V_{DD} = 0.3 V, 0.5 V, and 0.7 V as well as those of Si-CMOS with $W_{n\text{-CMOS}}$ = 32 nm and $W_{n\text{-CMOS}}$ = 42 nm [1].

Figure 7.4 shows the waveform of a GNRFET inverter for various roughness amplitudes and that of Si-CMOS at the scaled supply voltage of V_{DD} = 0.5 V. While the smooth-edge GNRFET has steeper transition than Si-CMOS, increasing the roughness amplitude results in higher propagation delay due to the smaller transistor current in diffusive and localization regimes and thereby the waveform is deteriorated correspondingly.

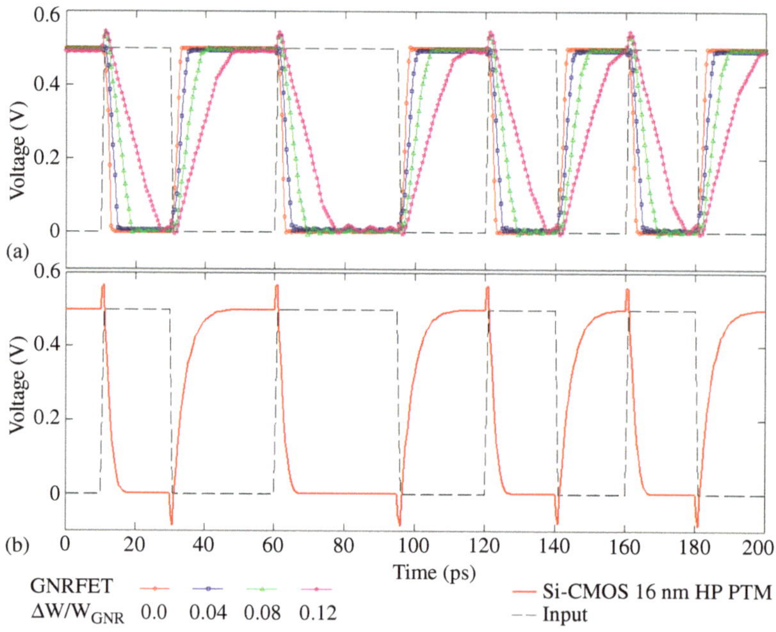

Figure 7.4 (a) Waveform of a GNRFET inverter with GNR(15,0) channel for various roughness amplitudes and that of (b) Si-CMOS at the scaled supply voltage of V_{DD} = 0.5 V [1].

7.4 Power and Delay of GNRFET Circuits

To evaluate the delay and power dissipation of GNRFETs for various LER amplitudes, a 5-stage fanout-of-4 buffer chain is implemented, as shown in Fig. 7.5a. This circuit is a standard benchmark for evaluating the process-independent delay metric, in which each inverter stage is derived by an inverter four times smaller than itself, while it drives an inverter four times larger than itself (corresponding to N_p in the figure). The average delay, static power, dynamic power, total power dissipation, and EDP of benchmark circuits for GNRFET and Si-CMOS are investigated by scaling the supply voltages and GNR widths.

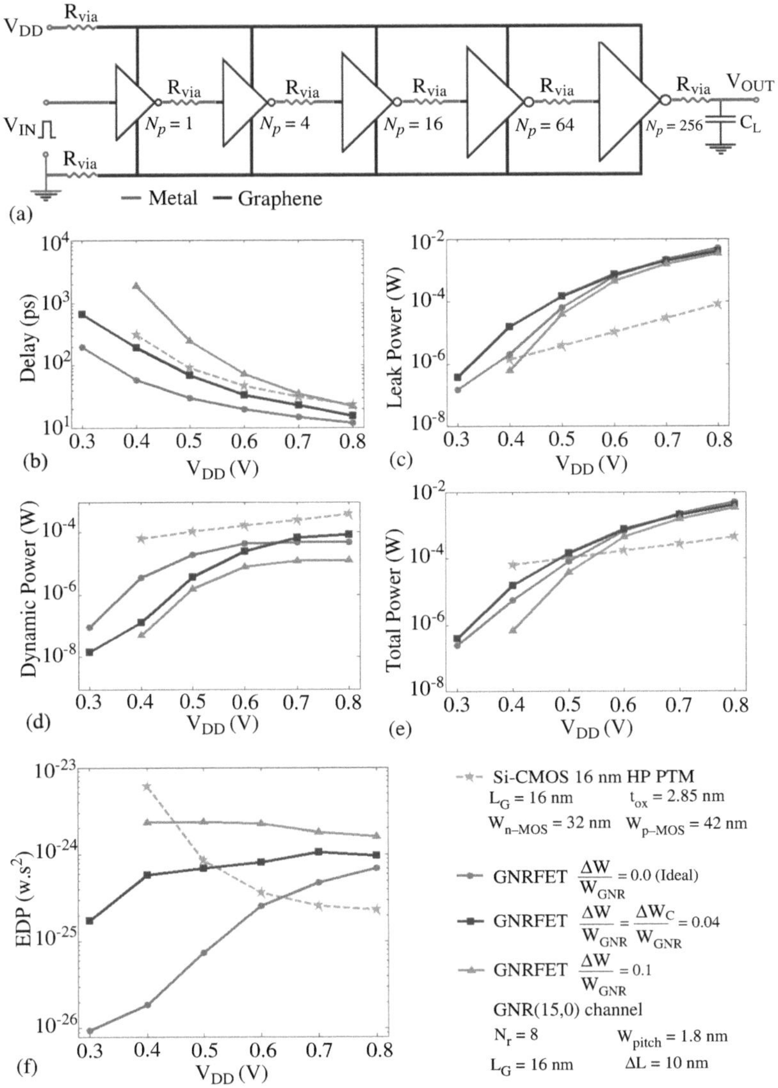

Figure 7.5 (a) A 5-stage fanout-of-4 buffer chain with R_{via} = 20 kΩ and C_L = 10 aF. (b) Delay, (c) leak power, (d) dynamic power, (e) total power, and (f) energy-delay product (EDP) of Si-CMOS and GNRFETs with ideal and rough edges versus supply voltages, V_{DD} [1].

Figure 7.5b shows that the delay of ideal-edge GNRFETs is increased from 12 ps to 200 ps by scaling the supply voltages from 0.8 V to 0.3 V, while that of Si-CMOS increases from 22 ps to 300 ps. The delay of GNRFETs with rough edges is larger than ideal-edge GNR. The rough-edge GNRFETs in the localization regime ($\Delta W/W_{GNR}$ = 0.1) show a larger trend of increasing delay with voltage scaling due to the smaller drive current. Thus, the delay of GNRFETs with highly rough edges can be the most limiting effects in scaling down the supply voltage. For instance, the GNRFET circuit shows six times larger delay than Si-CMOS at the scaled supply voltage of V_{DD} = 0.4 V, while they are approximately same at V_{DD} = 0.8 V. As shown in Fig. 7.5c, the leakage power is decreased by scaling the supply voltage as the possibility of BTBT from drain to channel is significantly reduced by scaling the supply voltage. While the leakage current of ideal GNRFET is approximately 40 times larger than Si-CMOS at V_{DD} = 0.7 V due to the smaller bandgap of the GNR(15,0) channel, both devices show the very close leak power ($\cong$ 2 μW) at the scaled supply voltage of V_{DD} = 0.4 V. For the roughness amplitude of $\Delta W/W_{GNR}$ = 0.04, however, this leak power increases to 16 μW due to the contribution of edge states in the BTBT of carriers. In the graphs, the values of GNRFETs with the roughness amplitude of $\Delta W/W_{GNR}$ = 0.1 and Si-CMOS are not reported at V_{DD} = 0.3 V as their output waveforms are completely deteriorated at these scaled supply voltages.

Figure 7.5d shows the dynamic power dissipation of Si-CMOS and GNRFETs versus scaled supply voltages. CMOS technology shows higher dynamic power dissipation than GNRFETs, especially at scaled supply voltages. In general, the effective channel area under the gate of Si-CMOS is larger than multi-channel GNRFETs, resulting in larger input capacitance and corresponding dynamic power. Also both the effective mass and the bandgap of GNR(15,0) are smaller than those of silicon, which leads to higher ON-state current at the scaled supply voltage. This promises the superior dynamic performance of GNRFET circuits by scaling the supply voltage. Figure 7.5e shows the total power dissipation of GNRFETs with smooth and rough edges in comparison with CMOS technology. It can be seen that ideal-edge

GNRFETs and GNRFETs with roughness amplitude of $\Delta W/W_{GNR}$ = 0.04 can operate at V_{DD} = 0.3 V, consuming 0.25 μW and 0.0.4 μW, respectively, which are 250 × and 160 × lower than that of Si-CMOS (at V_{DD} = 0.4 V).

The EDP can reveal the tradeoffs between delay and power dissipation of GNRFETs and Si-CMOS circuits. It can be seen in Fig. 7.5f that the EDP of Si-CMOS is significantly increased by scaling down the supply voltages, while that of ideal GNRFETs is decreased, such that GNRFETs show 320 × lower EDP at V_{DD} = 0.4 V. The ITRS has predicted that such high-mobility materials can continue the improvement in switching speed at the same time with much lower switching power consumption [10]. However, EDP is increased by increasing the roughness amplitudes and it cannot be reduced by scaling down the supply voltage as the mobility is limited by carrier scattering due to LER. For instance, at V_{DD} = 0.4 V, the 320 × lower EDP of ideal GNRFETs than Si-CMOS becomes 10% and 40% of Si-CMOS for $\Delta W/W_{GNR}$ = 0.04 and $\Delta W/W_{GNR}$ = 0.1, respectively.

Figure 7.6 shows the effect of GNR width (or GNR index, N) on the delay and power dissipation of GNRFETs at V_{DD} = 0.5 V. In other words, the supply voltage is now fixed, and the bandgap varies inversely with the GNR width to investigate the importance of bandgap engineering and the effects of edge roughness on GNRFETs. By increasing the GNR width, the bandgap is reduced and the upper sub-bands can get populated and have more contribution to carrier transport in GNRFETs. This increases the OFF-state and ON-state currents, leading to smaller delay ($\tau = CV/I$) and larger leak power, as shown in Fig. 7.6a and 7.6b, respectively. While the delay increases continuously by increasing the LER amplitudes, the leak power reduces by increasing the edge scattering after the initial increase due to enhanced BTBT. Figure 7.6c shows that the dynamic powers of the circuit implemented with ideal-edge GNRs are almost constant with GNR width, while its dependence on GNR width increases for non-ideal GNRFETs. As shown in Fig. 7.6d, the total power dissipation is increased with the same trend as leak power because it is the dominant power-consuming process by decreasing the bandgap of GNR channels. The EDP of rough-edge GNRFETs is decreased by increasing the GNR width and shows larger values than ideal-edge GNRFETs, as shown in Fig. 7.6e.

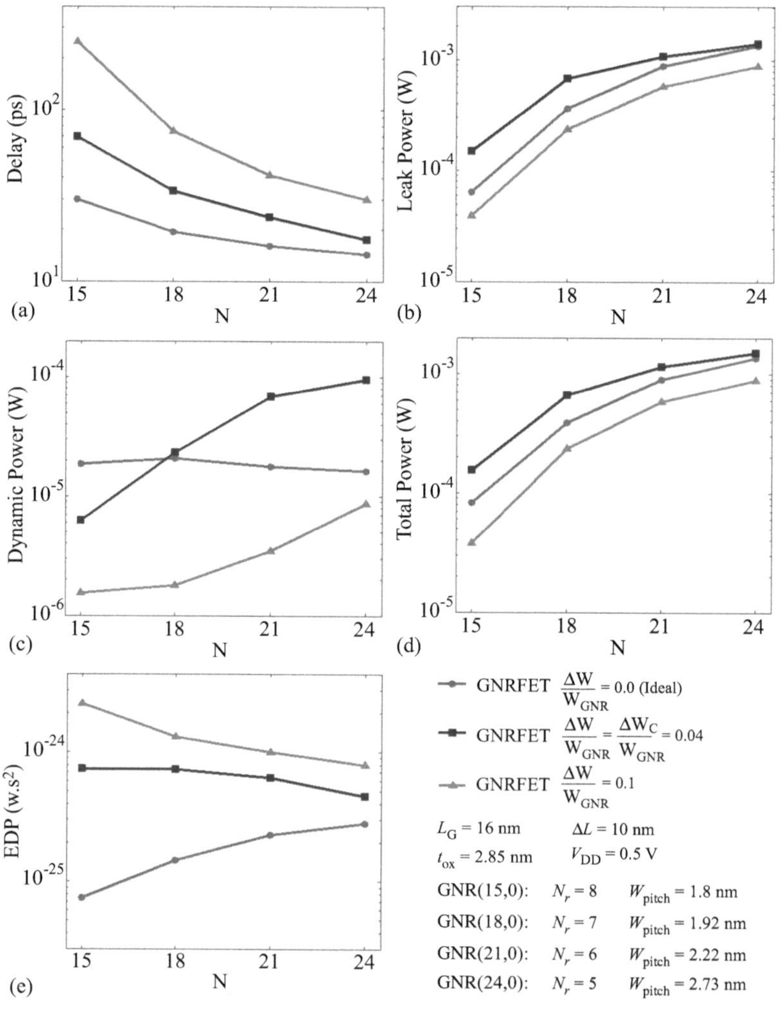

Figure 7.6 (a) Delay, (b) leak power, (c) dynamic power, (d) total power, and (e) energy-delay product (EDP) of GNRFETs with ideal and rough edges of GNR(N,0) channels versus GNR index, N. *Note*: The number of GNR channels (N_r) and the pitch spacing between ribbons (W_{pitch}) are presented in the figure, which results in the same physical width [1].

7.5 GNRFET-Based Energy Recovery Logic Design

Although aggressive scaling of transistor dimensions and increasing chip complexity have satisfied the demand for increasing the performance of integrated circuits, reducing the power dissipation has been always a major research effort for applications in portable devices. At the device level, power dissipation has been reduced mainly by scaling down the supply voltage and reducing capacitances with the scaling of transistors and interconnects. However, transistor and interconnect dimensions are limited in size due to the physics behind it. Furthermore, small geometry effects in MOSFETs [11] play a major role in the performance of integrated circuits as these increase the leakage current and consequently the static power dissipation. The well-known Moore's law is nearly approaching to an end in the next decade; there is increasing efforts in search of new materials such as carbon nanotubes (CNTs) [12] and graphene [11, 13] possibly substituting silicon in integrated circuits. The ITRS has predicted that the higher carrier mobilities in these materials allow more aggressive supply voltage scaling at the same time with higher drive current than in silicon MOSFETs, resulting in improvement in both speed and power dissipation [10]. Graphene is a promising alternative to silicon due to its atomically thin planar structure, high carrier concentration, high carrier mobilities, and thermal conductivity [14]. Patterning narrow stripes of graphene known as GNRs can open a bandgap of several hundred meV due to the quantum confinement of carriers [15] in order to make GNRFETs suitable for logic applications. At the circuit level, energy loss or heat dissipation can be reduced by energy recovery techniques [16] using the two key rules of clocked adiabatic logic: Never turn on a transistor when there is a potential difference across it, and never turn off a transistor when current is flowing through it. Novel GNRFET-based circuits can be desired for ultra-low energy operation.

XOR and XNOR gates play an important role in the ALU of core processors [17, 18]. XOR/XNOR gates have been implemented in energy recovery CMOS circuits [19] and in emerging technologies based on CNTs [20]. In this chapter, we extend our design of

XOR/XNOR gates to a new clocked adiabatic logic using an emerging 2D material (graphene)-based FETs. The proposed GNRFET-based XOR/XNOR gates can work with the scaled supply voltage V_{DD} down to 0.7 V at 45 nm technology node. It can reduce the energy dissipation and power density below the reported value in the ITRS [10]. This section presents several CAL XOR/XNOR circuits and the proposed GNRFET-based CAL XOR/XNOR circuits.

In this section, we use a GNRFET structure with armchair chirality GNR(13,0) as the channel material in MOSFET-like structure. The nanoribbon under the gate is intrinsic, while the extension regions between the gate regions and the source and drain contacts are doped with the concentration of 0.001 n-type dopants per carbon atom to form Ohmic contacts. The insulator layer is aluminum nitride (AlN) with the relative dielectric permittivity $\kappa = 9$ and thickness $t_{ox} = 1$ nm, which can be deposited on graphene with good uniformity and reproducibility [21]. The length and width of the channel are $L_G = 45$ nm and $W_G = 5.0$ nm, respectively. The supply voltage V_{DD} is 1.1 V for 45 nm technology node, which has been scaled down to 0.7 V to show the low-voltage operation of GNRFETs for decreasing power density in the proposed XOR/XNOR gates.

We have simulated different energy-efficient GNRFET-based XOR/XNOR circuits as done in conventional CMOS circuits [12]. Figure 7.7a shows 2N-2N2P XNOR/XOR circuit [22] as a modified version of efficient charge recovery logic (ECRL) and stated under partial energy recovery logic family [16]. In the circuit, the cross-coupled inverters latch the output, and the required logics of XNOR = (A + B)•(A + B) and XOR = (A•B) + (A•B) are then implemented by two complementary networks. The cross-coupled inverters stabilize the output logic, and two complementary pull-down networks increase the speed of discharging the inverters tree during the evaluation phase. Figure 7.7b shows the CAL [19, 23], where a clock signal controls the transistors M1 and M2 allowing the inverter circuit to operate with a single clocked power supply, Φ, and hold the output logic value corresponding to the previous evaluation phase when the clock signal disconnects the input logic network from the cross-coupled inverters [23]. Adding M3 and M4 in Fig. 7.7c can insulate the input part from the logic output part in order to cancel the leakage path of eight transistors of input networks. The complementary input logic networks are called the powerless (PL) XOR and ground-less (GL) XNOR circuits [17]. PL-GL CAL XOR/XNOR circuits transfer the basic disadvantage of their inputs to outputs, such that these display bad output levels and poor delay characteristics. These circuits cannot function properly by scaling down the supply voltage [17].

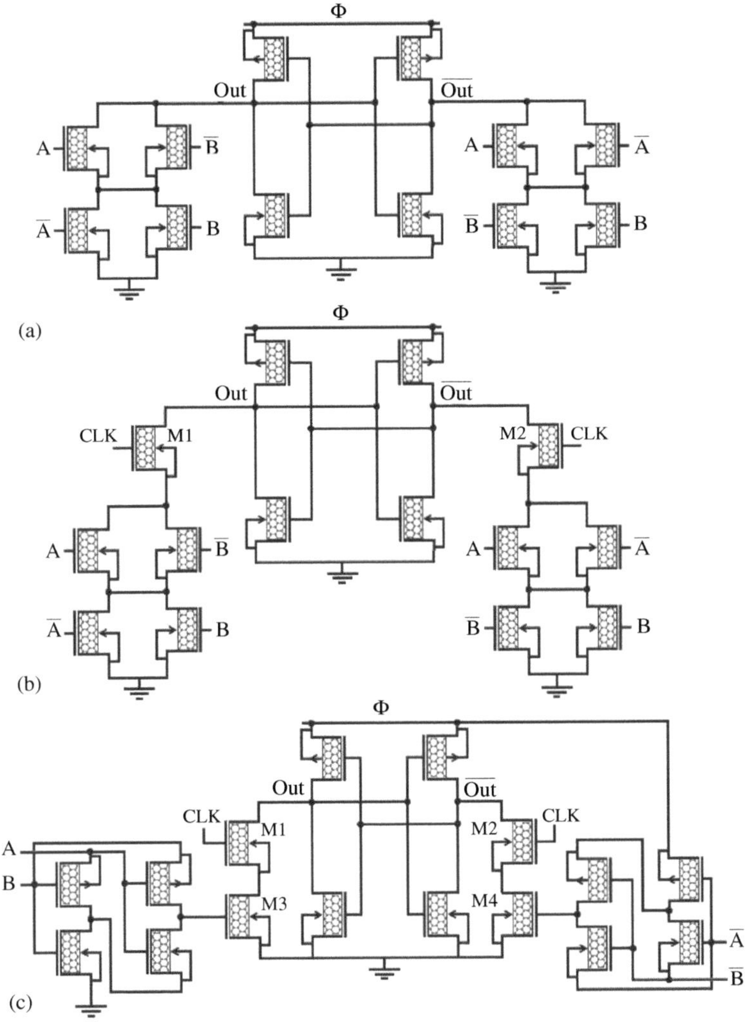

Figure 7.7 Circuit schematics of (a) 2N-2N2P XNOR/XOR, (b) basic CAL XOR/XNOR, and (c) PL-GL CAL XOR/XNOR. Reprinted from Ref. [24], Copyright 2015, with permission from IEEE.

Figure 7.8 shows our modification in GNRFET-based CAL XOR/XNOR, which consumes less power than the other CAL circuits [19, 24]. The implementation of both the input logic networks is combined in one circuit using only six transistors [25]. Likewise in the input circuits of Fig. 7.7c, there are a single connection to a clocked power supply, Φ, and a single connection to the ground with no direct path between them, which avoids the short-circuit currents while allowing the scaling of supply voltage. In addition, the input network does not need complementary of inputs A and B in the circuits of Fig. 7.7 and has two transistors less than in Fig. 7.7c for reduced power consumption.

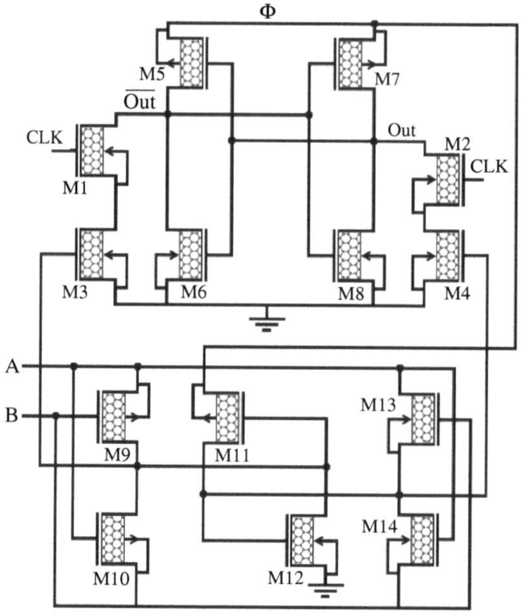

Figure 7.8 Circuit schematic of the new CAL XOR/XNOR gate. Reprinted from Ref. [24], Copyright 2015, with permission from IEEE.

Figure 7.9 shows the input and output waveforms of the proposed CAL XOR/XNOR. Switching the power clock, Φ has the trapezoid waveform with the same duration as the input waveforms A and B. In the evaluation phase, the clock signal CLK enables the logic evaluation and then the cross-coupled inverters restore the clock signal in latch configuration. In the hold phase, the CLK disables the logic evaluation, and the previously stored logic state repeats at outputs regardless of the current input signals. The circuit can operate at the scaled supply voltage V_{DD} equal to 0.8 V.

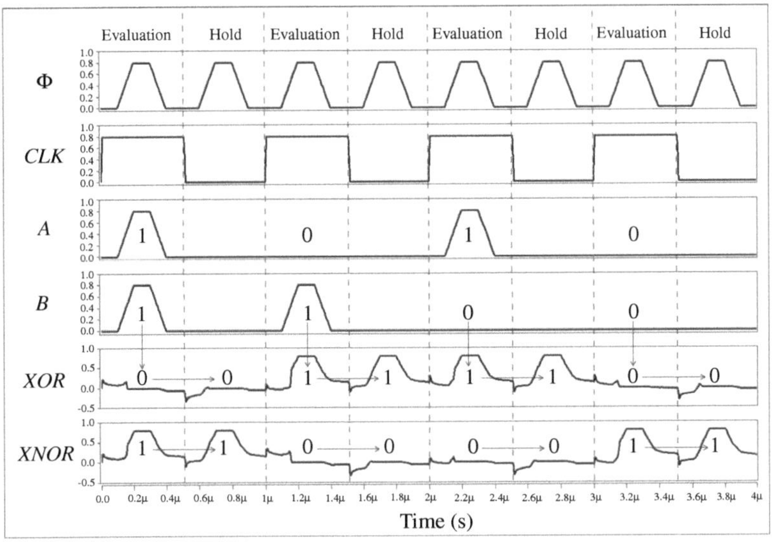

Figure 7.9 Input and output waveforms of the new CAL XOR/XNOR circuit. Reprinted from Ref. [24], Copyright 2015, with permission from IEEE.

The possibility of bandgap engineering allows to reduce the threshold voltage of GNRFET corresponding to the bandgap implied by the GNR width. Thus, there is another degree of freedom beside the channel length for GNRFET-based circuits, such that the proposed GNRFET circuits can operate at a much lower supply voltage. In conventional MOSFETs, the bandgap of silicon is fixed, and the threshold voltage and the corresponding supply voltage depend on the choice of the gate electrode material, dielectric constant, and thickness of oxide layer and substrate doping. The threshold voltage in GNRFETs can be tuned by the bandgap engineering of GNR such that wide GNRFETs can operate under scaled supply voltages. Thus, high-mobility materials such as GNR in channel can reduce the dynamic power consumption (CV_{DD}^2f) resulting in net improvement in both speed and power dissipation, as reported by the ITRS [10]. In addition, the effective mass of electrons and holes is same in GNRFETs such that complementary GNRFETs can operate at scaled voltages without the need for increasing the physical channel width in the pull-up network like conventional CMOS logic design.

The power density on a chip is a key consideration for down-scaling of VLSI circuits. The power density P_D can be calculated by $P_D = P/\alpha A$ [16], where P is the total power consumption, A is the active area of transistors, and α is the area factor, which accounts for the chip area in routing. We have assumed $\alpha = 4$ [16] in our simulations. Figure 7.10 shows the power densities of the XOR/XNOR circuits of Fig. 7.7 and our new CAL XOR/XNOR circuit in Fig. 7.8 at different frequencies for the supply voltage 1.1 V of the 45 nm technology node. The power densities of all the circuits are increased by increasing the operation frequency, while the new CAL XOR/XNOR has the least power density and decreases the power density by nearly 65% compared with the PL-GL CAL XOR/XNOR circuit. However, the power density at 1.1 V for the 45 nm technology node is still about 20 times higher than the limit in the ITRS [10]. The advantage of a GNRFET that has a high-mobility channel can provide some relief by allowing more aggressive V_{DD} scaling [10].

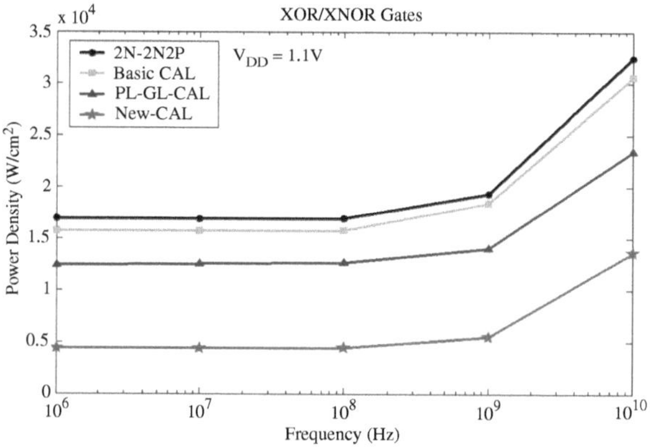

Figure 7.10 Power density versus frequency for different XOR/XNOR circuits. Reprinted from Ref. [24], Copyright 2015, with permission from IEEE.

Figure 7.11 shows the reduction in power densities of the ERL circuits by scaling down the supply voltage from 1.1 V to 0.7 V. The new CAL XOR/XNOR circuit can consume much less power and can satisfy the tolerable power density up to about 1 GHz for $V_{DD} = 0.7$ V. However, the circuit cannot operate in frequencies higher than 1 GHz for $V_{DD} = 0.7$ V, such that the circuit designers must use higher supply voltage for higher frequencies while keeping the power density within the acceptable level.

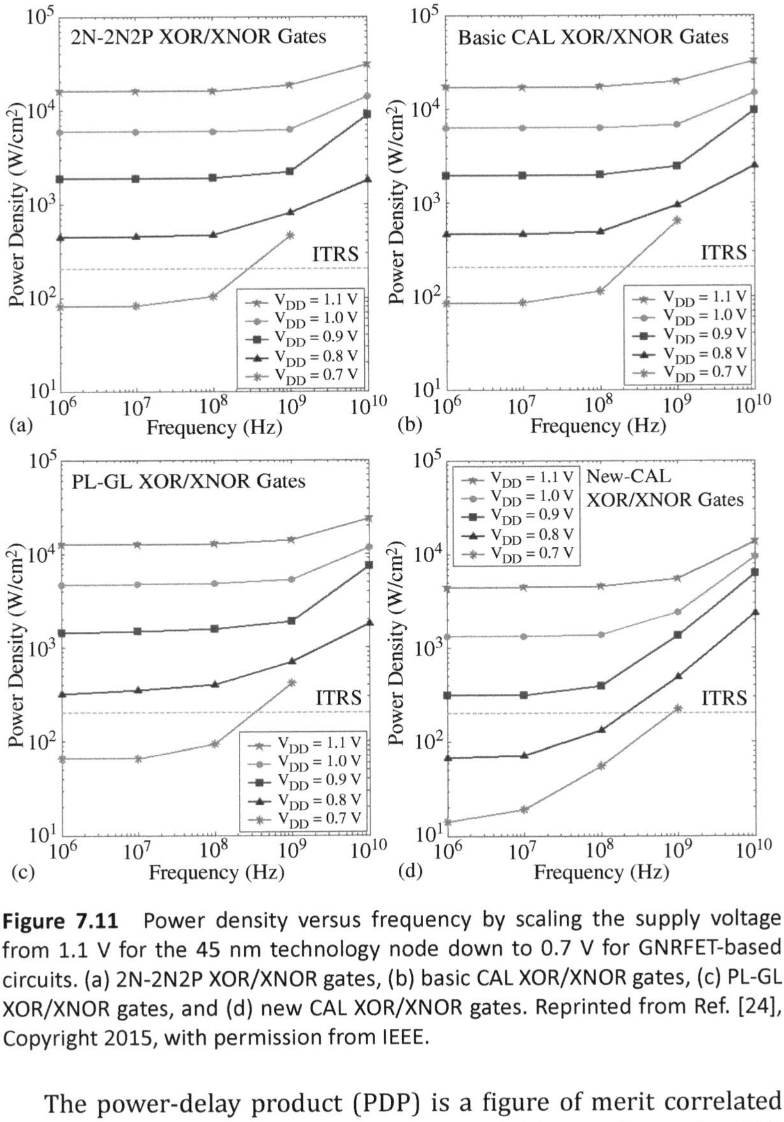

Figure 7.11 Power density versus frequency by scaling the supply voltage from 1.1 V for the 45 nm technology node down to 0.7 V for GNRFET-based circuits. (a) 2N-2N2P XOR/XNOR gates, (b) basic CAL XOR/XNOR gates, (c) PL-GL XOR/XNOR gates, and (d) new CAL XOR/XNOR gates. Reprinted from Ref. [24], Copyright 2015, with permission from IEEE.

The power-delay product (PDP) is a figure of merit correlated with the energy efficiency of a logic gate. Table 7.1 shows the PDP of PL-GL and the new CAL XOR/XNOR gates based on GNRFETs and CMOS at two different supply voltages and operation frequencies. The CMOS implementation used 45 nm high-performance libraries from the Predictive Technology Model [26] for the same device geometries as GNRFETs. In Table 7.1, comparing two rows shows the effectiveness of the proposed CAL XOR/XNOR gates in reducing the PDP, while comparing two columns shows the importance of GNRFET for low-power integrated circuit design. For V_{DD} = 0.7 V and f = 3 MHz, it can be seen that the CMOS implementation of PL-GL XOR/XNOR gates shows PDP equal to 440 × 10^{-18} W·s, while its implementation with GNRFET reduces the PDP to 300 × 10^{-18} W·s, leading to ~33% reduction. However, the major PDP reduction can be obtained by changing the design to the proposed new CAL XOR/XNOR gates, which results in approximately seven times reduction in PDP to 44 × 10^{-18} W·s. At a higher frequency of f = 60 MHz, we have to increase V_{DD} to 0.9 V as the CMOS cannot operate at the scaled supply voltage of 0.7 V. For V_{DD} = 0.9 V and f = 60 MHz, the same power reduction can be observed due to both the modification in the CAL circuit design and the implementation of the circuit using GNRFET. At V_{DD} = 0.9 V and f = 60 MHz, the GNRFET implementation of new CAL has ~60% more PDP than the operation at V_{DD} = 0.7 V and f = 3 MHz, while its CMOS implementation shows approximately four times increase in PDP, indicating the importance of GNRFET for low-voltage and high-frequency operation.

Table 7.1 Power-delay product (PDP) of PL-GL and the new CAL XOR/XNOR gates for the implementations based on GNRFETs and conventional CMOS at V_{DD} = 0.7 V, f = 3 MHz and V_{DD} = 0.9 V, f = 60 MHz

	PDP (10^{18} W·s) at V_{DD} = 0.7 V and f = 3 MHz		**PDP (10^{18} W·s) at V_{DD} = 0.7 V and f = 3 MHz**	
	GNRFET	**CMOS**	**GNRFET**	**CMOS**
PL-GL XOR/ XNOR GATES	300	440	420	680
NEW CAL XOR/ XNOR GATES	44	56	70	220

Source: Ref. [24].

7.6 Summary

In this chapter, we have studied the analytical model of multi-channel GNRFETs to investigate the effect of edge roughness on the power and delay performance of GNRFETs and the advantage of replacing conventional MOSFET with GNRFETs in power-efficient logic designs. Using the SPICE circuit simulation, we showed that increasing the GNR width deteriorates the OFF-state performance of GNRFETs, such that narrower armchair GNRs improve the device robustness to short-channel effects, leading to better OFF-state performance considering smaller OFF-current, larger I_{ON}/I_{OFF} ratio, smaller subthreshold swing, and smaller drain-induced barrier lowering. The wider armchair GNRs allow the scaling of channel length and supply voltage resulting in better ON-state performance such as higher drive current, smaller intrinsic gate-delay time, and smaller power-delay product. In addition, the width-dependent characteristics of GNRFETs are investigated for two GNR semiconducting families (3p,0) and (3p+1,0). We find that GNRs(3p+1,0) demonstrate superior OFF-state performance, while, on the other hand, GNRs(3p,0) show superior ON-state performance. Thus, GNRs(3p+1,0) are promising for low-power design, while GNRs(3p,0) indicate a preferable attribute for high-frequency applications.

The GNRFET circuit model is also used for designing new GNRFET-based XNOR/XOR gates based on energy recovery techniques. We have showed that the proposed circuit can decrease the power density by nearly 65% compared to the earlier reported energy-efficient CAL XNOR/XOR circuits. Furthermore, we have showed that the supply voltage of GNRFET-based circuits can be scaled down, resulting in about three times less power density and remains below the accepted power density limit up to 1 GHz and 0.7 V supply voltage.

References

1. Banadaki, Y. M., Sharifi, S., Craig III, W. O., and Hou, H.-C. (2016). Power and delay performance of graphene-based circuits including edge roughness effects, *American Journal of Engineering Research*, **5**, pp. 266–277.
2. Berger, C., Song, Z., Li, X., Wu, X., Brown, N., Naud, C., Mayou, D., Li, T., Hass, J., and Marchenkov, A. N. (2006). Electronic confinement and coherence in patterned epitaxial graphene, *Science*, **312**, pp. 1191–1196.
3. Van Noorden, R. (2006). Moving towards a graphene world, *Nature*, **442**, pp. 228–229.

4. Banadaki, Y. M. and Srivastava, A. (2015). Investigation of the width-dependent static characteristics of graphene nanoribbon field effect transistors using non-parabolic quantum-based model, *Solid-State Electronics*, **111**, pp. 80–90.
5. Kang, J., Sarkar, D., Khatami, Y., and Banerjee, K. (2013). Proposal for all-graphene monolithic logic circuits, *Applied Physics Letters*, **103**, p. 083113.
6. Nagashio, K., Nishimura, T., Kita, K., and Toriumi, A. (2009) Metal/graphene contact as a performance killer of ultra-high mobility graphene analysis of intrinsic mobility and contact resistance, *2009 IEEE International Electron Devices Meeting (IEDM)*, Baltimore, USA, pp. 1–4.
7. Chen, Z., Lin, Y.-M., Rooks, M. J., and Avouris, P. (2007). Graphene nano-ribbon electronics, *Physica E: Low-dimensional Systems and Nanostructures*, **40**, pp. 228–232.
8. Han, M. Y., Özyilmaz, B., Zhang, Y., and Kim, P. (2007). Energy band-gap engineering of graphene nanoribbons, *Physical Review Letters*, **98**, p. 206805.
9. Choudhury, M. R., Yoon, Y., Guo, J., and Mohanram, K. (2011). Graphene nanoribbon FETs: Technology exploration for performance and reliability, *IEEE Transactions on Nanotechnology*, **10**, pp. 727–736.
10. Wilson, L. (2013). *Semiconductor Industry Association*, International Technology Roadmap for Semiconductors (ITRS).
11. Banadaki, Y. M. and Srivastava, A. (2016). Effect of edge roughness on static characteristics of graphene nanoribbon field effect transistor, *Electronics*, **5**, p. 11.
12. Srivastava, A., Marulanda, J. M., Xu, Y., and Sharma, A. (2015). *Carbon-Based Electronics: Transistors and Interconnects at the Nanoscale*, CRC Press.
13. Banadaki, Y. M. and Srivastava, A. (2015). Investigation of the width-dependent static characteristics of graphene nanoribbon field effect transistors using non-parabolic quantum-based model, *Solid-State Electronics*, **112**, pp. 1–9.
14. Cooper, D. R., D'Anjou, B., Ghattamaneni, N., Harack, B., Hilke, M., Horth, A., Majlis, N., Massicotte, M., Vandsburger, L., and Whiteway, E. (2012). Experimental review of graphene, *International Scholarly Research Notices*, **2012**, p. 501686.
15. Datta, S. (2005). *Quantum Transport: Atom to Transistor*, Cambridge University Press.

16. Hwang, M.-E., Raychowdhury, A., and Roy, K. (2005). Energy-recovery techniques to reduce on-chip power density in molecular nanotechnologies, *IEEE Transactions on Circuits and Systems I: Regular Papers*, **52**, pp. 1580–1589.

17. Goel, S., Elgamel, M., and Bayoumi, M. A. (2003). Novel design methodology for high-performance XOR-XNOR circuit design, *16th Symposium on Integrated Circuits and Systems Design, 2003. SBCCI 2003. Proceedings*, Sao Paulo, Brazil, pp. 71–76.

18. Wang, J.-M., Fang, S.-C., and Feng, W.-S. (1994). New efficient designs for XOR and XNOR functions on the transistor level, *IEEE Journal of Solid-State Circuits*, **29**, pp. 780–786.

19. Xu, Y. and Srivastava, A. (2007). New energy recovery CMOS XNOR/XOR gates, *2007 50th Midwest Symposium on Circuits and Systems*, Montreal, Canada, pp. 948–951.

20. Srivastava, A., Xu, Y., and Soundararajan, R. (2010). Energy recovery techniques for CNT-FET circuits, *2010 International Symposium on Electronic System Design*, Bhubaneswar, India, pp. 81–84.

21. Owlia, H. and Keshavarzi, P. (2014). Investigation of the novel attributes of a double-gate graphene nanoribbon FET with AlN high-κ dielectrics, *Superlattices and Microstructures*, **75**, pp. 613–620.

22. Kramer, A., Denker, J. S., Flower, B., and Moroney, J. (1995). 2nd order adiabatic computation with 2N-2P and 2N-2N2P logic circuits, *Proceedings of the 1995 International Symposium on Low Power Design*, California, USA, pp. 191–196.

23. Maksimovic, D. and Oklobdzija, V. G. (1995). Clocked CMOS adiabatic logic with single AC power supply, *ESSCIRC '95: Twenty-first European Solid-State Circuits Conference*, Lille, France, pp. 370–373.

24. Banadaki, Y. M., Srivastava, A., and Sharifi, S. (2015). Clocked adiabatic XOR and XNOR CMOS gates design based on graphene nanoribbon complementary field effect transistors, *2015 IEEE International Symposium on Nanoelectronic and Information Systems*, Indore, India, pp. 13–18.

25. Radhakrishnan, D. (2001). Low-voltage low-power CMOS full adder, *IEE Proceedings: Circuits, Devices and Systems*, **148**, pp. 19–24.

26. Zhao, W. and Cao, Y. (2006). New generation of predictive technology model for sub-45 nm early design exploration, *IEEE Transactions on Electron Devices*, **53**, pp. 2816–2823.

Chapter 8

Graphene Sensing and Energy Recovery

8.1 Introduction

Aggressive scaling of transistor dimensions and increasing chip complexity have satisfied the demand for increasing the performance of integrated circuits for decades. The supply voltage and device capacitances can be scaled down by scaling down the device dimension. However, small-geometry effects play a major role in the performance degradation of integrated circuits as they increases the leakage current and consequently the static power dissipation [1]. Thus, the power dissipation and the corresponding die temperature have been continuously increased. Graphene has atomically thin planar structure, high carrier concentration, and high carrier mobility and thermal conductivity [2]. Patterning narrow stripes of graphene known as graphene nanoribbons (GNRs) can open a bandgap of several hundred mV due to the quantum confinement of carriers in order to make graphene nanoribbons field-effect transistor (GNRFET) suitable for logic applications [3]. GNR is not only a promising alternative channel material in very large scale integration (VLSI) technology, but also its atomistic thickness provides the maximum possible surface-to-bulk ratio, enabling its application in high-performance on-chip temperature sensors with nanometer dimensions for future integrated circuits. In recent years, unique properties of graphene have been exploited in

Graphene Nanostructures: Modeling, Simulation, and Applications in Electronics and Photonics
Yaser M. Banadaki and Safura Sharifi

ISBN 978-981-4800-36-5 (Hardcover), 978-0-429-02221-0 (eBook)
www.jennystanford.com

designing numerous sensors, including biosensors [4], gas sensors [8], chemical sensors [6], as well as ultrafast detectors [7] and temperature sensors [8, 9].

Graphene is also one of the ideal candidates for next-generation energy conversion and storage [10]. It is biologically non-invasive, light, flexible, transparent, and mechanically strong, which can convert the generated body heat into an electric current at room temperature [11], which would otherwise be wasted. The progressive evolution of carrier distribution function from the low-temperature on-source side to the high-temperature region on the drain side creates a thermal gradient along the GNR and thereby the thermoelectric current. In this chapter, we have also explored the feasibility of GNRs as a temperature sensor for detecting local on-chip temperature and a thermoelectric material for harvesting the dissipated heat in integrated circuits.

8.2 GNRFET-Based Temperature Sensors

Figure 8.1 shows the proposed GNRFET temperature sensor made up of multiple armchair GNRs as temperature-sensing material [11]. In the structure, the insulator layer is thin aluminum nitride (AlN), and a highly doped silicon substrate is a back-gate contact. The local heat generated due to high power density in nanometer size can transport through the top area or the substrate underneath the graphene sheet. The heat modulates the conductance of GNRs, which can be measured by correlating the temperature to the variation in its resistance. The device is a miniaturized temperature sensor with high sensitivity, low power consumption, and fast response time due to its high electrical and thermal conductivity [2].

The area of the temperature-sensing surface is equal to the gate length of GNRFET, L_G, multiplied by its width $W_G = N_{CH}(W_{GNR} + W_{sp}/2)$, where N_{CH} is the number of GNR channels and W_{sp} is the spacing between two GNRs. $W_{GNR} = \sqrt{3}(N+1)\,a_{cc}/2$ is the width of each GNR(N,0), where N is the armchair GNR index and a_{cc} is the carbon–carbon bonding distance. The gate length, channel spacing, and number of GNR channels are assumed to be L_G = 22 nm, W_{sp} = 2 nm, and N_{CH} = 15, respectively. In this study, the GNR index N is changed from 7 to 88 with an increment of 9 as a simulation parameter corresponding to the GNR width ranging from ~0.7 nm to ~10.8 nm. The relative dielectric permittivity of the AlN insulator layer is $\kappa = 9$, and its thickness is $t_{ox} = 1$ nm. The symmetric regions of the GNR channel between the gate and contacts are doped with the n-type dopants with a concentration of $f_{dop} = 5 \times 10^{-3}$ per carbon atom as the source and drain reservoirs. The nominal value of the supply voltage V_{DD} is 1 V for the 22 nm technology node.

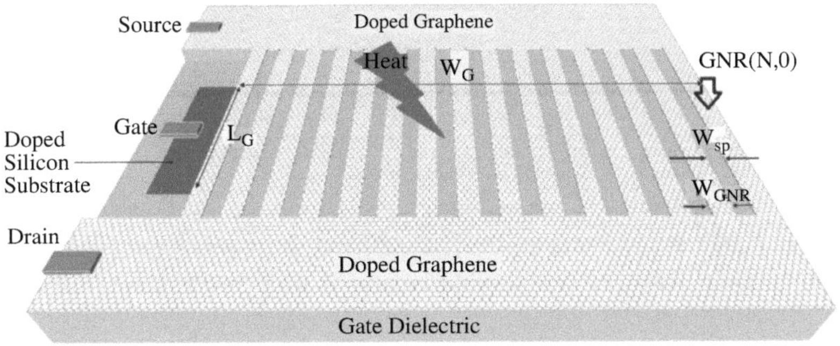

Figure 8.1 Schematic of a back-gate multi-channel GNRFET temperature sensor [11].

Figure 8.2 shows the change in GNRFET current by increasing the temperature for various gate biases. Increasing the temperature increases the current, contrary to the electrical resistance of a conductor such as a copper wire, where its current decreases on increasing the temperature due to the increase in carrier collisions in the material. For the 22-nm-length armchair GNRs as the channel material, there are bandgap and discrete conducting modes due to the quantum confinement of carriers, in which carriers can transport with very small collisions since the mean free path of carriers in GNRs is very large, reaching up to the μm range for smooth-edge GNRs. Thus, the heat imparted to the graphene surface partly increases the scattering rate by collision of carriers, but mostly increases the energy of carriers, allowing them to get populated in the upper sub-bands and consequently increasing the current. In low bias voltage (V_G = 0.3 V), the upper sub-bands have relatively small carriers and thus increasing the temperature significantly increases the population of carriers in the upper sub-bands. As such, increasing the temperature leads to a steeper increase in current at the scaled voltage of V_G = 0.3 V than at higher gate biases (e.g., V_G = 0.9 V).

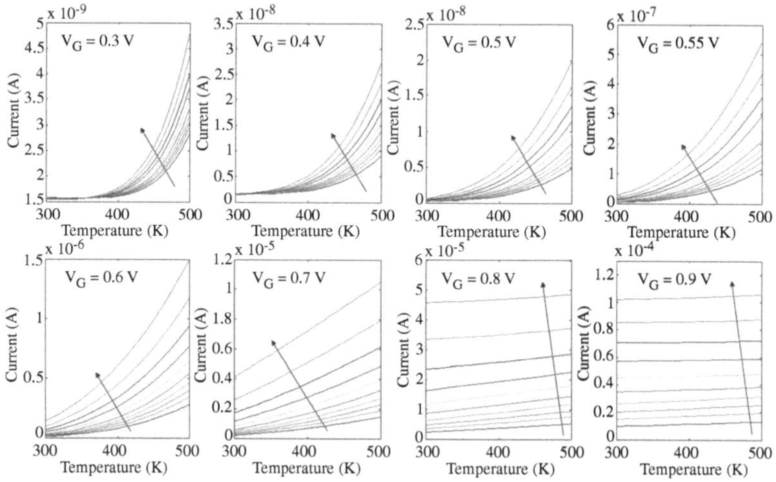

Figure 8.2 Change in GNRFET current by increasing the temperature for various gate biases. Arrows show the increase in GNR index *N* from 7 to 88 [11].

The temperature coefficient of resistance (TCR) of the proposed GNRFET-based nanometer temperature sensor has been calculated using $TCR = \Delta R/R_0\ \Delta T$, which represents the resistance change per degree of temperature variation. Figure 8.3 shows the TCR versus variation in sensing temperature at room-temperature set point for various GNR indices and gate biases. The maximum TCR depends on the accurate choosing of GNR width and gate bias conditions. At V_{GS} = 0.3 V, the maximum TCR belongs to the wider GNR(88,0), equal to 6×10^{-3}/K at 450 K (Fig. 8.3a), while it increases to 8.5×10^{-3}/K at 350 K for V_{GS} = 0.4 V (Fig. 8.3b). By increasing the gate bias to 0.5 V, the TCR of GNR(88,0) maximizes to 2.1×10^{-2}/K at room temperature (Fig. 8.3c). For V_{GS} = 0.55 V (Fig. 8.3d), the same maximum TCR occurred at room temperature for all GNRs, such that the TCR is almost independent of the GNR width, which is inversely related to the generated bandgap. While the maximum TCR of wider GNRs was larger for gate bias smaller than 0.55 V, the trend changes for larger gate biases and the GNRs with smaller widths have the maximum TCR at room temperature (Fig. 8.3e). By increasing the gate bias from 0.7 V to 0.8 V, and then to 0.9 V (Figs. 8.3f, 8.3g, and 8.3h), the maximum TCR of all GNRs decreases and the difference between them increases corresponding to the increase in GNR width dependence of TCR. These TCR plots reveal that the best gate bias voltage for the proposed GNRFET-based temperature sensor is V_{GS} = 0.55 V, in which the TCR maximizes at room temperature equal to 2.1×10^{-2}/K, and also it is independent of the GNR width, allowing its design with dimension-free W_G.

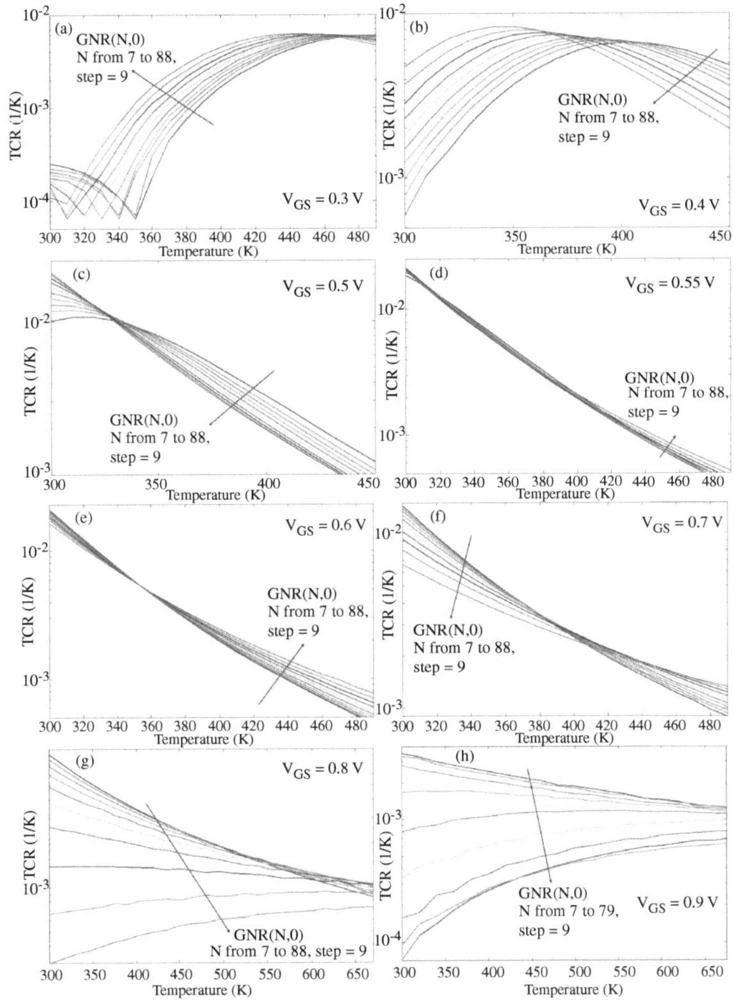

Figure 8.3 TCR versus variation in sensing temperature at room-temperature set point for various GNR indices and gate biases [11].

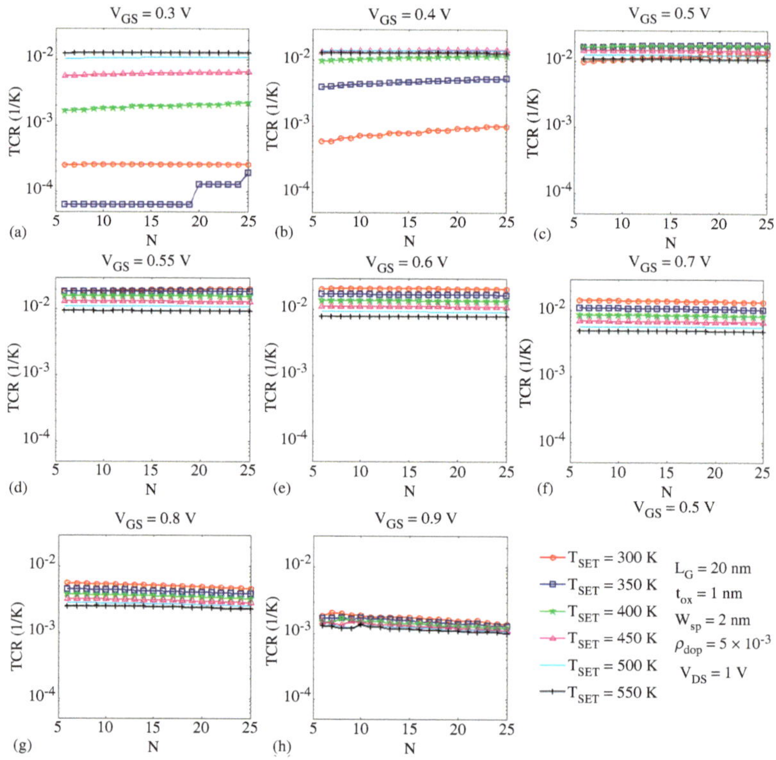

Figure 8.4 TCR versus GNR indices from 7 to 25 for various set point temperatures and gate bias conditions [11].

Figure 8.4 shows the TCR versus GNR indices from 6 to 25 for various set point temperatures and gate bias conditions. It can be observed from Fig. 8.4a that the GNRFET biased at a low gate voltage of 0.3 V shows high TCR at high-temperature set points, T_{set} = 550 K close to 1.2×10^{-2}/K, while that of room temperature is roughly two orders of magnitude smaller due to the lower number of sub-bands involved in the carrier transport. By increasing the gate bias voltage to 0.4 V and 0.5 V (Figs. 8.4b and 8.4c), the TCRs of all set points increase and come close together, which maximizes to ~2 $\times 10^{-2}$/K at 0.5 V, showing the same maximum values for the gate voltages of 0.55 V and 0.6 V (Figs. 8.4d and 8.4e). After that the maximum TCR of all temperature set points decreases by increasing the gate bias, such that they decrease an order of magnitude at V_{GS} = 0.9 V (Fig. 8.4h). Thus, a GNRFET-based temperature sensor with nanoscale dimensions demonstrates maximum sensitivity at scaled gate voltages close to half of the nominal supply voltage of the 22 nm technology node. In addition, note that the TCR of high-temperature set points, T_{set}, comes below the TCR of low-temperature set points by increasing the gate bias. As such, if the gate voltage is chosen above $V_G = 0.55$ V, the temperature close to room temperature can be sensed with higher sensitivity (larger TCR) than higher temperatures close to 550 K.

8.3 GNRFET for Energy Harvesting

Thermoelectricity is the direct conversion of temperature gradient to electrical potential difference. Despite its early discovery in the 1800s, there have always been efforts to find new materials for thermoelectric applications. Such materials can be used for power generation in various applications such as energy harvesting in automotives and integrated circuits, as well as efficient thermal-to-electrical energy conversion in wearable electronics. Graphene, a two-dimensional sheet of graphite, is one of the ideal candidates for next-generation electronics as well as energy conversion and storage [10] because of maximum possible surface-to-bulk ratio along with its exceptional electrical and thermal transports [12, 13]. For efficient energy generation from heat, the materials must have large thermoelectric figure-of-merit described by $ZT \cong (S^2\sigma T)/K$, where S is the Seebeck coefficient, σ is the electric conductance, T is the temperature, and K is the thermal conductivity. This means large Seebeck coefficient and electrical conductivity, and small thermal conductivity. However, graphene is an excellent heat conductor due to covalent bonding between carbon atoms, and patterned graphene might be a better candidate for thermoelectric application.

By pattering graphene in the form of GNRs, the carriers are confined to a quasi-one-dimensional transport system, resulting in opening up of bandgaps in graphene [2, 11]. GNRs are classified as zigzag and armchair depending on the edge geometry in the transport direction. Zigzag-edge GNRs are metallic in nature, whereas armchair-edge GNRs can be classified into three different families depending on the number of carbon atoms, N, in the transverse direction: the metallic family of $N = 3p+2$ and two semiconducting families of $N = 3p$ and $3p+1$, where p is an arbitrary integer [3]. This electron confinement quantizes the 2D wavevectors of graphene into sub-band energies depending on the ribbon width [1]. The possibility of tuning the bandgaps by controlling the width and edge geometry of graphene ribbons makes GNRs potentially useful for various electronic and thermoelectric applications. Despite high thermal conductivity of large-area graphene [14], it is predicted that the thermal conduction depends on the chiral angle and width of nanoribbons and the minimum value occurs in the armchair direction and for narrow nanoribbons [10]. As such, the carrier transport model is required to evaluate and optimize the figure-of-merit *ZT* of GNR for thermoelectric applications [15]. We investigate the thermoelectric properties of GNRs by computing the Seebeck coefficient, electrical conductivity, thermoelectric power (TEP) factor, and the electrical component of thermal conductivity and finally study the thermoelectric figure-of-merit *ZT* as a function of GNR width and chemical potential.

8.3.1 Thermoelectric Model

Electron transport and the associated thermoelectric properties can be derived from the semi-classical Boltzmann transport equation with the relaxation-time approximation using the differential conductivity equation as follows:

$$\sigma(E)=e^2\tau(E)v(E)^2\rho_{\mathrm{DOS}}(E)\left(-\frac{\partial f_0}{\partial E}\right) \tag{8.1}$$

where $\tau(E)$ is the total electron relaxation time, $v(E)$ is the electron velocity, $\rho_{\mathrm{DOS}}(E)$ is the density of states, and $f_0(E)$ is the Fermi–Dirac distribution. The total electron relaxation time is determined by relevant scattering mechanisms in GNR such as intrinsic acoustic phonons (APs) and optical phonons (OPs) [16], charged impurity [17], and line-edge roughness (LER) in narrow GNRs [18, 19]. Unlike large-area graphene, line-edge scattering is usually a dominant scattering mechanism in narrow GNRs. As such, decreasing the GNR width reduces the mean free path of carriers, as shown in Table 8.1. The electrical conductivity can be obtained by an integral of the differential conductivity over energy as $\sigma = F_0(E_F)$, where E_F is the Fermi level relative to the conduction band minimum and F_0 is the integral function as follows [20]:

$$F_m(E_F) = \int_0^\infty \sigma_d(E)(E - E_F)^m dE \tag{8.2}$$

The Seebeck coefficient is given by [20]

$$S = \frac{1}{eT}\frac{F_1(E_F)}{F_0(E_F)} \tag{8.3}$$

The electronic component of thermal conductivity is given by [20]

$$K_e = \frac{1}{e^2T}\left\{F_2(E_F) - \frac{[F_1(E_F)]^2}{F_0(E_F)}\right\} \tag{8.4}$$

The figure-of-merit *ZT* is the key parameter used to describe and compare the relative efficiency of thermoelectric materials, which is given by [21]

$$ZT = \frac{\sigma S^2 T}{K_l + K_e} \tag{8.5}$$

where *T* is the absolute temperature, K_e is the electrical component of thermal conductivity, and K_1 is the lattice component of thermal conductivity resulting from lattice vibrations (phonons) in solid. Recent measurements show an order of magnitude lower lattice thermal conductivity of GNR than that of micron-sized graphene at room temperature, ranging from 30 to 300 W/mK [22], as shown in Table 8.1.

Table 8.1 Minimum energy, effective mass, and non-parabolicity factor of GNRs used in the transport calculation

	E_{c1} (eV)	m_1*	α_1	E_{c2} (eV)	m_2*	α_2	λ_{mfp} (nm)	K_l
GNR(7,0)	0.76	0.215	0.9	1.05	0.155	0.7	2.7	30
GNR(13,0)	0.43	0.11	1.25	0.61	0.11	1.05	9.5	147
GNR(19,0)	0.30	0.07	1.8	0.43	0.08	1.3	20.5	188
GNR(25,0)	0.23	0.056	2.18	0.33	0.066	1.65	40.2	225
GNR(34,0)	0.17	0.037	3.3	0.25	0.066	0.073	79	270

Source: Ref. [3].

Note: The mean free path λ_{mfp} and lattice thermal conductivity K_l are obtained from Refs. [23] and [24], respectively, at charge concentration of $n_{2D} = 10^{12}$ cm^{-2} and room temperature.

The thermoelectric transport properties of graphene have been studied theoretically and experimentally [25–27]. In order to evaluate the TEP, a controlled temperature ΔT on a 10 nm long GNR is considered such that the temperatures at the source and drain ends are given by $T_S = 300$ and $T_D = 300 + \Delta T$, respectively. The small temperature difference $\Delta T = 10$ K is assumed to result in different carrier distribution function at two contacts and consequently the thermoelectric current. The thermoelectric characteristics are shown in terms of chemical potential shift in the Fermi energy in different GNR(N,0) channels as an important parameter for evaluating the electrical and thermal characteristics.

8.3.2 Electrical Conductivity

Figure 8.5 shows the electrical conductivity $\sigma = ne\mu$ versus the chemical potential for five GNRs in this study. Shifting the Fermi energy toward the conduction (valence) sub-bands increases the electron (hole) carriers in the GNR sub-bands and thereby increases the electrical conductivity. The ambipolar transport is partially recovered with regard to the generated bandgap of GNR around the charge neutrality point (CNP) similar to large-area graphene [28]. The increase in GNR width results in smaller bandgap and increases the carrier density at smaller energies.

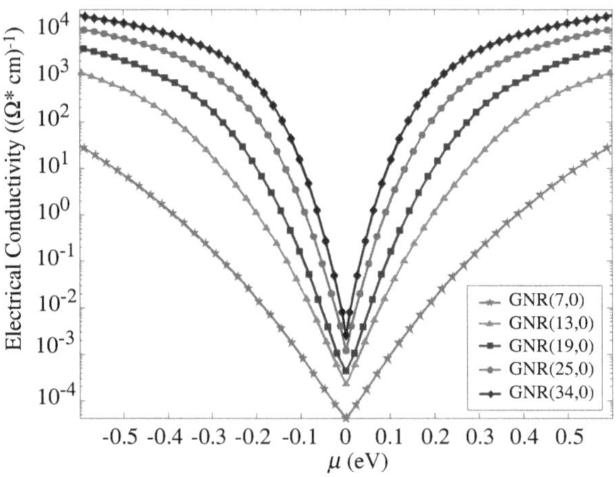

Figure 8.5 Electrical conductivity versus chemical potential for different GNRs [13].

8.3.3 Seebeck Coefficient

The Seebeck effect is the conversion of temperature difference into electrical voltage as $S = \Delta V/\Delta T$, where S is the Seebeck coefficient, ΔV is the measured electrical voltage difference, and ΔT is the temperature difference or gradient between the hot and the cold sides of a thermoelectric device. Figure 8.6 shows the Seebeck effect versus the chemical potential for five GNRs in this study. The TEP curve (or Seebeck coefficient) increases with the increase in the chemical potential close to CNP. However, it monotonically decreases by increasing the carrier concentration according to Mott's relation as $\partial\sigma/\partial E$ rolls off [26]. As such, the symmetry of the TEP curve can be explained by the symmetric conductance of GNR around the CNP in Fig. 8.5. The sign of the thermoelectric curve changes from positive to negative at CNP corresponding to the change in the type of majority charge carriers. The TEP curve of GNRs has a peak located close to the first sub-band energy. By decreasing the GNR width, the bandgap increases and thereby the maximum of TEP occurs at larger chemical potentials. However, the value of the TEP peak increases by decreasing the GNR width as well. For instance, the maximum TEP value of GNR(34,0) is calculated equal to 830 μV/K while that of GNR(7,0) is increased to 1170 μV/K at room temperature. The higher effective mass of GNR(7,0) produces a higher Seebeck coefficient due to the slower motion of heavier carriers. In other words, the

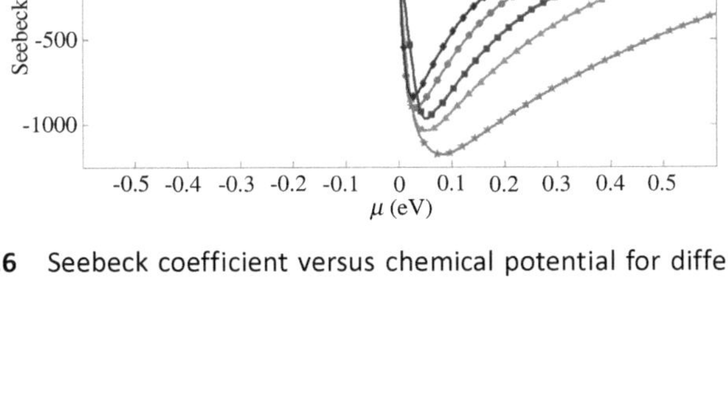

Figure 8.6 Seebeck coefficient versus chemical potential for different GNRs [13].

location of the first sub-band energy determines the TEP peak in energy, while the curvature of the sub-band energy determines the magnitude of the maximum TEP. This strengthens our conclusion that narrower armchair GNRs have properties that could be used to develop high-performance thermoelectric devices.

8.3.4 Electrical Thermal Conductivity

Figure 8.7 shows the electrical thermal conductivity K_e of GNRs as a function of chemical potentials for different GNRs. It can be observed that the electronic component of thermal conductivity is also increased by increasing the carrier concentration as the larger numbers of carriers can contribute to bring back the device to thermal equilibrium condition. Similarly, the wider GNR(34,0) shows larger electrical thermal conductivity because it has a larger number of sub-bands and consequently larger carriers can contribute to thermal conduction for a chemical potential. The electrical thermal conductivity has small contribution to the total thermal conductivity as the lattice thermal conductivity (see Table 8.1) is the dominant phenomenon for heat conduction in carbon materials. For instance, the electrical thermal conductivity of GNR(34,0) at a large chemical potential of 0.6 eV can reach to about 28 W/(mK), which is approximately 10% of its lattice thermal conductivity. The strong covalent sp_2 bonding in graphene results in efficient heat transfer by lattice vibrations at room temperature [29]. If the lattice component of thermal conductivity is effectively suppressed by AP confinement and scattering by LER and defects, the thermoelectric efficiency of GNR-based devices can be enhanced [29]. Reducing the lattice thermal conduction requires the knowledge of phonon dispersion and is beyond this work.

8.3.5 Power Factor

The Seebeck coefficient (S) and the electrical conductivity (σ) are used to obtain the power factor ($P = S^2\sigma$), as shown in Fig. 8.8. The power factor is a numerator in the figure-of-merit ZT relation, and thus an important quantity for maximizing the TEP of GNRs. The power factor exhibits the minimum value for the Fermi level close to the CNP as large values of the Seebeck coefficient are eliminated by the very small electrical conductivity of GNR in this region. By increasing the chemical potential, the power factor increases rapidly at the vicinity of the first sub-band of GNRs. However, further increase in chemical potential decreases the trend of the power factor as the decrease in Seebeck coefficient dominates the increase in the electrical conductance. As such, it is expected that the wider GNRs exhibit the peak in power factor in correspondence with Ref. [30]. In general, the larger power factor can be obtained for wider GNRs. For instance, at μ = 0.3 eV, the TEP of GNR(19,0) is measured equal to 10 μW/(cmK^2), while it increases to 23 μW/(cmK^2) and 38 μW/(cmK^2) for GNR(25,0) and GNR(34,0), respectively.

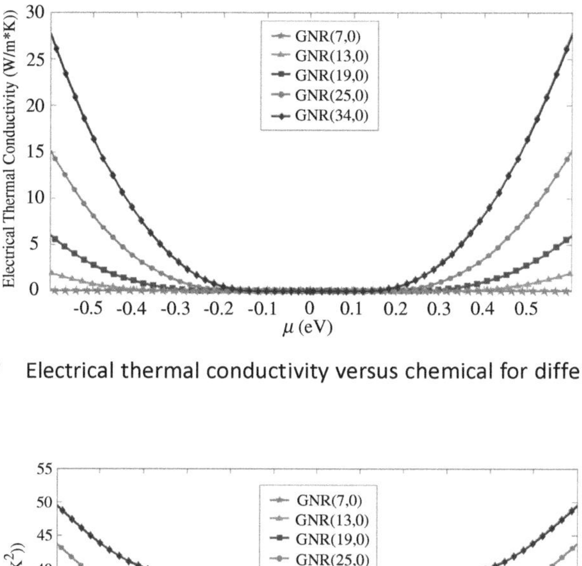

Figure 8.7 Electrical thermal conductivity versus chemical for different GNRs [13].

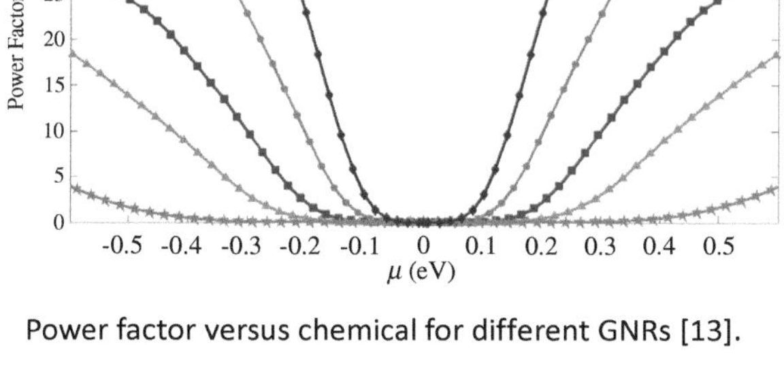

Figure 8.8 Power factor versus chemical for different GNRs [13].

8.3.6 Thermoelectric Figure-of-Merit *ZT*

The efficiency of GNRs as thermoelectric materials can be evaluated by the dimensionless thermoelectric figure-of-merit $ZT = \sigma S^2 T/(K_1 + K_e)$. Figure 8.9 shows the value of *ZT* as a function of chemical potential for different GNRs. The thermoelectric figure-of-merit *ZT* is close to zero at the vicinity of the CNP, such that the length of zero region in energy depends on the GNR bandgap. The *ZT* increases by increasing the chemical potential beyond the first sub-band of GNRs. The wider GNR(34,0) exhibits larger *ZT* value at a smaller chemical potential due to the smaller bandgap and effective mass, while the trend of the curve decreases by increasing the chemical potentials. As such, at μ = 0.5 eV, all the GNRs in the study show approximately the same value of *ZT* = 0.01. The calculated value is close to the measured values of common semiconductor materials such as Si and GaAs at room temperature, but it is much smaller than that of common thermoelectric materials such as BiTe and PbTe with *ZT*~0.4 at room temperature [31]. In general, the maximum efficiency of an ideal thermoelectric material can be obtained if it has the large electrical conductivity of a metal, the high Seebeck coefficient of a semiconductor, and the low thermal conductivity of glass. While GNRs can overcome the first two criteria, the primary cause of low *ZT* is the large lattice thermal conductivity. Further increase in the *ZT* value of GNR requires phonon engineering to reduce the thermal conductivity of graphene without a significant reduction in its TEP and electrical conductivity [32].

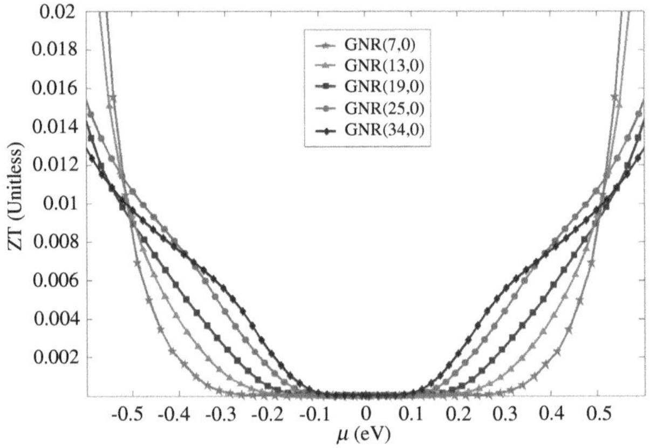

Figure 8.9 Thermoelectric figure-of-merit *ZT* versus chemical potential for different GNRs [13].

8.4 Summary

Graphene has been extensively investigated as a promising material for various types of high-performance sensors due to its large surface-to-volume ratio, remarkably high carrier mobility, high carrier density, high thermal conductivity, extremely high mechanical strength, and high signal-to-noise ratio. The power density and the corresponding die temperature can be tremendously high in scaled emerging technology designs, urging the on-chip sensing and controlling of the generated heat in nanometer dimensions.

In this chapter, we have explored the feasibility of a thin oxide GNR as a nanometer-size temperature sensor for detecting the local on-chip temperature at scaled bias voltages of emerging technology. We have introduced an analytical model of GNRFET for the 22 nm technology node, which incorporates both thermionic emission of high-energy carriers and band-to-band tunneling (BTBT) of carriers from drain to channel regions together with different scattering mechanisms such as intrinsic APs, OPs, and LER in narrow GNRs. The TCR of GNRFET-based temperature sensors shows approximately an order of magnitude higher TCR than large-area graphene FET temperature sensors by accurately choosing the GNR width and the bias condition for a temperature set point. At gate bias $V_{GS} = 0.55$ V, the TCR maximizes at room temperature to 2.1×10^{-2}/K, which is also independent of GNR width, allowing the design of width-free GNRFETs for room-temperature sensing applications.

The thermoelectric properties of GNRs have been evaluated by using the semi-classical Boltzmann theory in the relaxation-time approximation. The width dependence of thermoelectric characteristics has been studied by simulating five members of the armchair-edge GNR($3p$+1,0) family. We found that narrower GNRs exhibit larger Seebeck coefficient and smaller thermal conductivity. These properties potentially provide a path toward improved thermoelectric figure-of-merit ZT. However, narrower GNRs also show smaller electrical conductivity, which limits the thermoelectric performance at small chemical potentials. As such, wider GNRs exhibit larger ZT value at smaller chemical potential of Fermi levels, while narrower GNRs exhibit better thermoelectric performance by increasing the chemical potential. In general, the calculated values of ZT for all the GNRs are much smaller than common thermoelectric materials at room temperature. Changing the GNR width cannot be enough to optimize the thermoelectric performance of GNRs because of the tradeoff in thermal and electrical parameters. While the optimization of a parameter directly affects another parameter in an undesired way, GNRs with different edge geometry, roughness, and passivation must be studied to minimize the phonon heat transmission without a significant decrease in electrical carrier transport.

References

1. Banadaki, Y. M. and Srivastava, A. (2015). Scaling effects on static metrics and switching attributes of graphene nanoribbon FET for emerging technology, *IEEE Transactions on Emerging Topics in Computing,* **3**, pp. 458–469.
2. Cooper, D. R., D'Anjou, B., Ghattamaneni, N., Harack, B., Hilke, M., Horth, A., Majlis, N., Massicotte, M., Vandsburger, L., and Whiteway, E. (2012). Experimental review of graphene, *ISRN Condensed Matter Physics,* 2012, 501686, p. 56.
3. Banadaki, Y. M. and Srivastava, A. (2015). Investigation of the width-dependent static characteristics of graphene nanoribbon field effect transistors using non-parabolic quantum-based model, *Solid-State Electronics,* **111**, pp. 80–90.
4. Kuila, T., Bose, S., Khanra, P., K. Mishra, A., Kim, N. H., and Lee, J. H. (2011). Recent advances in graphene-based biosensors, *Biosensors and Bioelectronics,* **26**, pp. 4637–4648.
5. Rumyantsev, S., Liu, G., Shur, M. S., Potyrailo, R. A., and Balandin, A. A. (2012). Selective gas sensing with a single pristine graphene transistor, *Nano Letters,* **12**, pp. 2294–2298.
6. Gan, T. and Hu, S. (2011). Electrochemical sensors based on graphene materials, *Microchimica Acta,* **175**, p. 1.
7. Vicarelli, L., Vitiello, M. S., Coquillat, D., Lombardo, A., Ferrari, A. C., Knap, W., Knap, W., Polini, M., Pellegrini, V., and Tredicucci, A. (2012). Graphene field-effect transistors as room-temperature terahertz detectors, *Nature Materials,* **11**, p. 865.
8. Sahoo, S., Barik, S. K., Sharma, G. L., Khurana, G., Scott, J. F., and Katiyar, R. S. (2012). Reduced graphene oxide as ultra fast temperature sensor. arXiv preprint arXiv:1204.1928.

9. Banadaki, Y., Mohsin, K., and Srivastava, A. (2014). A graphene field effect transistor for high temperature sensing applications, Proc. SPIE 9060, *Nanosensors, Biosensors, and Info-Tech Sensors and Systems*, p. 90600F.

10. Choi, H.-J., Jung, S.-M., Seo, J.-M., Chang, D. W., Dai, L., and Baek, J.-B. (2012). Graphene for energy conversion and storage in fuel cells and supercapacitors, *Nano Energy*, **1**, pp. 534–551.

11. Banadaki, Y. M., Srivastava, A., and Sharifi, S. (2016). Graphene nanoribbon field effect transistor for nanometer-size on-chip temperature sensor, *Proceedings of the SPIE*, **9802**, p. 980203.

12. Huang, B., Li, Z., Liu, Z., Zhou, G., Hao, S., Wu, J., and Duan, W. (2008). Adsorption of gas molecules on graphene nanoribbons and its implication for nanoscale molecule sensor. *The Journal of Physical Chemistry C*, **112**, pp.13442–13446.

13. Banadaki, Y. M., Hou, H.-C., and Sharifi, S. (2017). Graphene field effect transistor for generating on-chip thermoelectric power, *Proc. SPIE 10167, Nanosensors, Biosensors, Info-Tech Sensors and 3D Systems*, 101671I.

14. Balandin, A. A., Ghosh, S., Bao, W., Calizo, I., Teweldebrhan, D., Miao, F., and Lau, C. N. (2008). Superior thermal conductivity of single-layer graphene, *Nano Letters*, **8**, pp. 902–907

15. Sevinçli, H. and Cuniberti, G. (2010). Enhanced thermoelectric figure of merit in edge-disordered zigzag graphene nanoribbons, *Physical Review B*, **81**, p. 113401.

16. Akturk, A. and Goldsman, N. (2008). Electron transport and full-band electron-phonon interactions in graphene, *Journal of Applied Physics*, **103**, p. 053702.

17. Chen, J.-H., Jang, C., Xiao, S., Ishigami, M., and Fuhrer, M. S. (2008). Intrinsic and extrinsic performance limits of graphene devices on SiO_2, *Nature Nanotechnology*, **3**, pp. 206–209.

18. Banadaki, Y. M. and Srivastava, A. (2016). Effect of edge roughness on static characteristics of graphene nanoribbon field effect transistor, *Electronics*, **5**, p. 11.

19. Banadaki, Y. M., Sharifi, S., Craig III, W. O., and Hou, H. C. (2016). Power and delay performance of graphene-based circuits including edge roughness effects. *American Journal of Engineering Research*, **5**, pp. 266–277.

20. Bahk, J.-H., Bian, Z., Zebarjadi, M., Zide, J. M., Lu, H., Xu, D., Feser, J. P., Zeng, G., Majumdar, A., and Gossard, A. C. (2010). Thermoelectric

figure of merit of (In0.53Ga0.47As)0.8(In0.52Al0.48As)0.2 III-V semiconductor alloys, *Physical Review B*, **81**, p. 235209.

21. Dollfus, P., Nguyen, V. H., and Saint-Martin, J. (2015). Thermoelectric effects in graphene nanostructures, *Journal of Physics: Condensed Matter*, **27**, p. 133204.
22. Bae, M.-H., Li, Z., Aksamija, Z., Martin, P. N., Xiong, F., Ong, Z.-Y., Knezevic, I., and Pop, E. (2013). Ballistic to diffusive crossover of heat flow in graphene ribbons, *Nature Communications*, **4**, pp. 1734.
23. Fiori, G. and Iannaccone, G. (2013). Multiscale modeling for graphene-based nanoscale transistors, *Proceedings of the IEEE*, **101**, pp. 1653–1669.
24. Wang, Y., Qiu, B., and Ruan, X. (2012). Edge effect on thermal transport in graphene nanoribbons: A phonon localization mechanism beyond edge roughness scattering, *Applied Physics Letters*, **101**, p. 013101.
25. Hwang, E., Rossi, E., and Sarma, S. D. (2009). Theory of thermopower in two-dimensional graphene, *Physical Review B*, **80**, p. 235415.
26. Zuev, Y. M., Chang, W., and Kim, P. (2009). Thermoelectric and magnetothermoelectric transport measurements of graphene, *Physical Review Letters*, **102**, p. 096807.
27. Löfwander, T. and Fogelström, M. (2007). Impurity scattering and Mott's formula in graphene, *Physical Review B*, **76**, p. 193401.
28. Lin, Y.-M., Perebeinos, V., Chen, Z., and Avouris, P. (2008). Electrical observation of subband formation in graphene nanoribbons, *Physical Review B*, **78**, p. 161409.
29. Nissimagoudar, A. and Sankeshwar, N. (2014). Significant reduction of lattice thermal conductivity due to phonon confinement in graphene nanoribbons, *Physical Review B*, **89**, p. 235422.
30. Reshak, A. (2015). Thermoelectric properties of fully hydrogenated graphene: Semi-classical Boltzmann theory, *Journal of Applied Physics*, **117**, p. 225104.
31. Snyder, G. J. and Toberer, E. S. (2008). Complex thermoelectric materials, *Nature Materials*, **7**, pp. 105–114.
32. Ghahari, F., Xie, H.-Y., Taniguchi, T., Watanabe, K., Foster, M. S., and Kim, P. (2016). Enhanced thermoelectric power in graphene: Violation of the Mott relation by inelastic scattering, *Physical Review Letters*, **116**, p. 136802.

Chapter 9

Graphene Photonic Properties and Applications

9.1 Introduction

Graphene is a one-atom-thick sheet of sp^2-hybridized carbon atoms arranged in a honeycomb lattice, and thereby the thinnest possible material in nature. Since its discovery, graphene has soon emerged as a promising nanomaterial for novel applications in electronics as well as in photonics due to its remarkable electrical and optical properties, which will be explained to some extent in this chapter. The discovery of graphene opened a new field of research wherein the thickness of the materials under investigation is orders of magnitude smaller than the wavelength of the light involved. The gapless nature of graphene attracted significant attention due to its strong interaction with photons in a wide energy range from microwave to ultraviolet [1, 2], as well as its high carrier mobility for high-speed applications in a broad wavelength range. The interesting property of graphene is its ability to absorb 2.3% of white light in spite of being only one-atom thick. The absorption coefficient of graphene exceeds 5×10^7 m^{-1} in the visible range if it is normalized to its atomic thickness, which is more than 10 times larger than those of gallium arsenide and silicon [3, 4]. This absorption is dictated by the fine structure constant rather than material specifics, so that the same amount of white light can be absorbed by adding each layer of

Graphene Nanostructures: Modeling, Simulation, and Applications in Electronics and Photonics
Yaser M. Banadaki and Safura Sharifi

ISBN 978-981-4800-36-5 (Hardcover), 978-0-429-02221-0 (eBook)
www.jennystanford.com

graphene. As such, the interaction with light can be further enhanced using graphene multilayers for practical device applications.

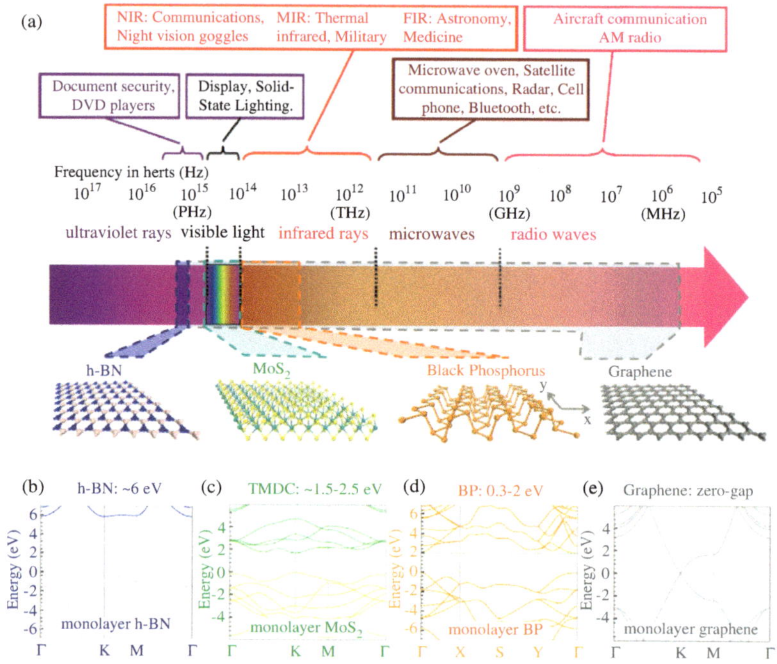

Figure 9.1 (a) A diagram of the electromagnetic spectrum represented by a rainbow arrow. Applications utilizing different spectral ranges are presented in the top portion of the panel. NIR, MIR, and FIR correspond to near, mid, and far infrared, respectively. The atomic structures of hexagonal boron nitride (h-BN), molybdenum disulfide (MoS_2), black phosphorus (BP), and graphene are shown in the bottom of the panel (from left to right). (b), (c), (d), (e) The band structures of a single layer of h-BN, MoS_2, BP, and graphene, respectively. Reprinted from Ref. [4], Copyright 2014, with permission from Springer Nature.

A wide range of two-dimensional (2D) materials makes it possible to construct various nanophotonic devices by combining them with different number of layers and compositions. Many 2D materials strongly interact with light and cover a very wide electromagnetic spectrum from microwave to ultraviolet due to their diverse electronic properties, as shown in Fig. 9.1 [4]. For instance, hexagonal boron nitride (h-BN) has 2D atomic structure similar to graphene, and it has been proposed as a promising complementary insulator layer [5], as shown in Fig. 9.1b. h-BN can be incorporated in various heterostructures for insulating other conducting 2D materials due to its large bandgap of ~6 eV. Single-layer transition metal dichalcogenides (TMDCs) such as MoS_2 and WSe_2 are also direct bandgap semiconductors [6, 7], as shown in Fig. 9.1c. These 2D materials have a sizable optical bandgap greater than 1 eV, which leads to many applications such as the realization of efficient light-emitting devices. Another recently discovered layered material with a direct bandgap is black phosphorus (BP) [8] (Fig. 9.1d). The bandgap of BP is sizable: from ~0.3 eV in its bulk form to ~2 eV by reducing the number of layers [9, 10], allowing the realization of mid-infrared photonic devices in a broad wavelength range.

9.2 Photonic Properties

A 2D graphene sheet can be physically regarded as one layer of 3D bulk graphite, but it possesses distinct properties for photonic applications compared to traditional semiconductors. The zero bandgap and linear dispersion of graphene result in very broadband strong absorptance. The linear dispersion of Dirac fermions implies that there will always be an electron–hole pair for broadband illumination, which is very different from semiconductors with bandgap and parabolic dispersion relations.

The optical absorption of suspended graphene has been measured to be ~2.3% across visible and near-infrared (NIR) frequencies [11]. The value is equal to the universal constant $\pi\alpha$, where $\alpha = e^2/(4\pi\varepsilon_0\hbar c) \approx 1/37$ is the fine structure constant [12]. The parameter describes coupling between light and relativistic electrons and is traditionally associated with quantum electrodynamics. The obtained value of optical absorption is remarkable for a one-atom-thick material, indicating that graphene can absorb a significant fraction of incident light.

Figure 9.2a shows the photograph of a 50 mm aperture partially covered by suspended graphene and its bilayer [11]. The line scan across the image shows the changes in the intensity of transmitted white light. Graphene's absorption is measured close to 2.3% so that the opacity is almost independent of the wavelength for visible illumination, as shown in Fig. 9.2b. The inset shows that increasing the number of graphene layers by one enhances the absorption of the structure by ~2.3%. As such, the optical absorption can be increased by stacking a few layers of graphene (FLG) with approximately the same ratio per layer of graphene. However, such structure needs to be isolated electrically by using dielectric insulators such as h-BN to prevent the formation of graphite.

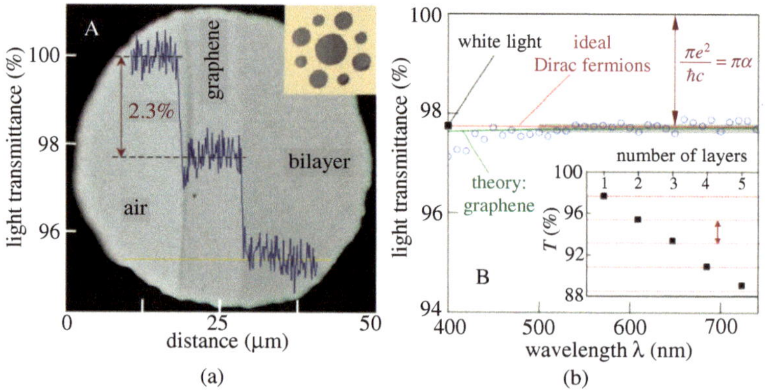

Figure 9.2 (a) Photograph of a 50 mm aperture partially covered by graphene and its bilayer. The line scan profile shows the intensity of the transmitted white light along the yellow line. The inset is a 20-μm-thick metal support structure with apertures of 20, 30, and 50 μm in diameter with graphene placed over them. (b) Transmittance spectrum of single-layer graphene (open circles). Slightly lower transmittance for λ < 500 nm is probably due to hydrocarbon contamination. The red line is the transmittance $T = (1 + 0.5\pi\alpha)^{-2}$ expected for 2D Dirac fermions, whereas the green curve takes into account the nonlinearity and triangular warping of graphene's electronic spectrum. The gray area indicates the standard error for measurements. The inset shows the transmittance of white light as a function of the number of graphene layers (squares). The dashed lines correspond to an intensity reduction by $\pi\alpha$ with each added layer. Adapted from Ref. [11], Copyright 2008, with permission from The American Association for the Advancement of Science.

9.3 Graphene Photonic Applications

As an atomic-thick material, graphene exhibits wavelength-independent absorption when the optical energy is below 3 eV, as well as complete transparency when the optical energy is smaller than double the Fermi level. These properties make graphene a promising candidate for many controllable photonic devices, as shown in Fig. 9.3. In this section, an overview of potential graphene photonics applications is presented together with the expected time for their realization.

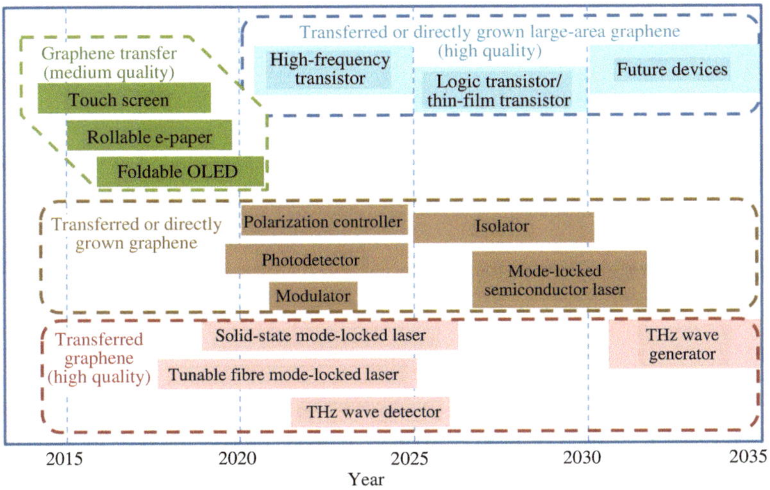

Figure 9.3 Graphene photonic applications in display technology (green), electronic devices (blue), as well as optical devices (pink), and interconnects (brown). Application timelines are based on projections of products requiring advanced materials such as graphene, which can be enabled by continuous advances in graphene technology. The figure gives an indication of when a functional device prototype could be expected based on device roadmaps and the development schedules of industry leaders. The quality of graphene and the required synthesis methods are shown for each category of application. The details are beyond the scope of this book. Reprinted from Ref. [20], Copyright 2012, with permission from Springer Nature.

9.3.1 Transparent Conductive Films and Passive Photonic Devices

Graphene is expected to substitute indium tin oxide (ITO) [13] in transparent conductive films used in devices such as light-emitting diodes (LEDs) and organic LEDs (OLEDs), solar cells, touchscreens, smart windows, and liquid crystal display (LCD) films [14, 15]. For these applications, graphene has superior characteristics such as high conductivity, mechanical flexibility, and optical transparency. Graphene not only provides larger transmittance than ITO (about 97% compared to 90%) but is also less expensive and less brittle than ITO. Recent research is focused on decreasing the sheet resistance of graphene by external doping to make it a competitive alternative for these applications [16]. Another application of graphene is for shielding electromagnetic radiation in microwave and terahertz ranges. At these frequencies, five layers of doped graphene can provide ~20 dB shielding efficiency [17], while millimeter thicknesses of traditional absorbers, such as ferrite or metal nanocomposites, are needed for such an effective shielding [18, 19]. As shown in Fig. 9.3, a graphene-based foldable touchscreen is expected to come into play in as early as 2020 [20].

9.3.2 Photodetectors

The spectral range and maximum bandwidth of traditional photodetectors based on silicon, germanium, or compound semiconductors are limited by the bandgap energy and carrier transit times of the materials [21]. On the other hand, due to the absence of a bandgap and the high carrier mobility of graphene, graphene-based photodetectors can, in principle, be used in a wide spectral range from ultraviolet to terahertz frequencies. Ultrafast extraction of photo-generated carriers is enabled by the high carrier mobility in graphene, resulting in extremely high bandwidth operation. As such, the transit-time-limited bandwidth of graphene photodetectors can possibly reach 1.5 THz [22], while that of InGaAs photodetectors is limited to 150 GHz [21]. It has been already demonstrated [22, 23] that infrared light pulses irradiated at the rate of 10 GHz and 40 GHz can be reliably detected by graphene–metal junctions, demonstrating the suitability of graphene for high-speed optical communications in the near future. However, the atomic thinness of graphene limits its absorption, so that the maximum responsivity of a graphene photodetector is low, much smaller than the required value of 1 AW^{-1} for effective detection [23].

There are several possible ways to increase the sensitivity of graphene photodetectors, such as using plasmonic nanostructures for the enhancement of the local optical electric field [24], integrating it with a waveguide to increase the light–graphene interaction length [25], or simply increasing the number of graphene layers in an aperiodic multilayer structure, as proposed in the current thesis. For instance, at the resonant wavelength, a 20-fold enhancement of the photocurrent has been achieved by incorporating a graphene layer inside Fabry–Pérot microcavities [26] or a ~60% absorption by utilizing Bragg reflectors [27]. It is expected that a graphene photodetector with a bandwidth of over 100 GHz will be competitive after 2020 [20], as shown in Fig. 9.3.

9.3.3 Optical Modulator

Optical modulators with high speed, small footprint, and high bandwidth are highly desirable for optical interconnects to encode transmission data by altering the light properties [28]. However, conventional electro-absorption modulators have limited operation bandwidth due to the slow switching time of silicon, which decreases their bandwidth to usually less than ~50 GHz [29]. Not only graphene absorption modulators have ultrafast response in absorbing a small amount of incident light over an ultrawide range of wavelengths, but also the interband transitions of photo-generated electrons can be modulated over broad spectral ranges by tuning the Fermi level of graphene layers [30]. Despite the strong light–graphene interaction, the absorption of a light beam by a single graphene layer is insufficient so that it is usually coupled with a silicon waveguide to increase the absorption in the NIR range [31]. The optical resonance of a photonic crystal nanocavity can be modulated by tuning the Fermi level in graphene layers, by reducing the resistance so as to reduce RC delay times, and by modifying the structures [32, 33], so that operation bandwidths in the terahertz range are expected. However, such developments are not expected before 2020, as shown in Fig. 9.3 [20].

9.3.4 Mode-Locked Laser

Graphene is also promising to find an application as a saturable absorber in fiber and laser technologies, wherein the absorption of light by a material decreases with increasing light intensity [34]. Saturated absorption can be observed in widely used semiconductors at very high optical intensities [35]; however, graphene reaches saturation at a lower intensity, by absorbing a significant amount of light energy per unit thickness [36, 37]. As such, graphene-saturable absorbers demonstrate ultrafast carrier relaxation time, controllable modulation depth, high damage threshold, high thermal conductivity, high stability, and wide tunability of the spectral absorption range [11, 38, 39]. Graphene-based tunable mode-locked lasers are expected to be commercialized by 2020 [20]. A graphene-saturable absorber could also enable an ultrafast passively mode-locked laser to operate at many different wavelengths for various applications such as spectroscopy, optical interconnects [40], bio-medicine [41], and material micromachining [42]. However, this application is expected to be demonstrated only after developing highly integrated optical interconnects around the late 2020s, as shown in Fig. 9.3 [20].

9.3.5 THz Wave Generator

THz wave generators are needed for various applications such as medical imaging, chemical sensors, and security devices. Graphene can be used as a gain medium to generate stimulated emission by optical pumping. However, as graphene has the same diffusion times for electrons and holes, the formation of a dipole may not be sufficient to overcome the stimulated emission threshold for continuous THz wave generation [43]. To generate waves in this range, it has been suggested to accelerate the carriers in a single layer or multiple layers of graphene using a femtosecond laser pulse field. However, due to the weak intensity of the generated waves in comparison with III–V semiconductors, it is expected that graphene-based THz wave generators are unlikely to emerge before 2030, as shown in Fig. 9.3 [20].

9.3.6 Optical Polarization Controller

Optical polarization controllers are crucial passive components to manipulate the polarization properties of photons. The excitation of Dirac fermions in graphene enables the differential attenuation of the transverse-magnetic mode, leading to an excellent extinction ratio of ~27 dB for broadband communications [44]. A hybrid graphene-based optical polarizer, in which a high-quality large-size graphene layer is integrated with an optical fiber, is expected to be fabricated by 2020 [20].

9.4 Optical Conductivity Model of Graphene

The refractive index is not well defined for 2D graphene because there is no rigorous definition for the induced polarization per unit volume. A more suitable physical quantity to describe the optical properties of graphene is the optical conductivity, also known as the carrier dynamic or AC conductivity of graphene. The optical conductivity is a complex number associated with the surface current induced in the 2D crystal by light [45]. The absorption is proportional to the real part of the optical conductivity [46], and the imaginary part is closely related to its behavior for application as a hyperbolic metamaterial. For instance, the sign of the imaginary part of the optical conductivity determines whether graphene supports transverse-electric or transverse-magnetic plasmons [47].

A large body of literature focuses on different techniques to model the complex conductivity of graphene [48–52]. This section introduces and reviews the analytical expression for the complex conductivity of graphene. The response of graphene's optical conductivity to an external electric field was derived theoretically by several researchers [49, 53–55]. Using the non-interacting linear response theory [45, 55], electrons are considered to move due to the applied electric field, which is the sum of the external field and the self-consistent field induced by all the electrons. The cone-like dispersion near Dirac points is assumed, since the exact band structure results in only a few percent difference in the conductivity at optical frequencies [45]. As a graphene sheet is an infinitesimally thin, non-local two-sided surface in free space in the x–y plane, one can obtain an extended non-local anisotropic model of conductivity using a magneto-optical surface complex conductivity tensor as follows [51, 56]:

$$\vec{\sigma}(\omega,\mu_c(\mathbf{E_0}),\Gamma,T,\mathbf{B_0}) = \hat{x}\hat{x}\sigma_{xx} + \hat{x}\hat{y}\sigma_{xy} + \hat{y}\hat{x}\sigma_{yx} + \hat{y}\hat{y}\sigma_{yy} \quad (9.1)$$

where ω is the radian frequency, μ_c is the chemical potential, Γ is the charge particle scattering rate, T is the temperature, and $\mathbf{E_0}$ and $\mathbf{B_0}$ are the DC electric and magnetic fields, respectively. The chemical potential can be controlled by chemical doping [57] or by applying DC bias field perpendicular to the graphene surface ($\mathbf{E_0} = \hat{z}E_0$) [58].

The geometry of graphene sheet implies that $\sigma_{xx} = \sigma_{yy}$ and $\sigma_{xy} = -\sigma_{yx}$, and thus Eq. (9.1) can be rewritten as follows:

$$\vec{\sigma}(\omega,\mu_c(\mathbf{E_0}),\Gamma,T,\mathbf{B_0})=\sigma_d\overleftrightarrow{\mathbf{I}_t}+\sigma_0\overleftrightarrow{\mathbf{J}_t} \tag{9.2}$$

where $\overleftrightarrow{\mathbf{I}_t}=\hat{x}\hat{x}+\hat{y}\hat{y}$ and $\overleftrightarrow{\mathbf{J}_t}=\hat{x}\hat{x}-\hat{y}\hat{y}$ are the symmetric and anti-symmetric dyads, respectively. σ_d and σ_0 are the diagonal and off-diagonal (Hall) conductivities, respectively, in the presence of electric potential and magnetic bias fields, which can be obtained using the Kubo formula [59, 60] as follows:

$$\begin{aligned}\sigma_d &= \frac{e^2 v_F^2 |eB_0| \hbar(\omega+i2\Gamma)}{i\pi} \\ &\times \sum_{n=0}^{\infty}\left\{\left(1-\frac{\Delta^2}{M_n M_{n+1}}\right)\frac{\left[n_F(M_n)-n_F(M_{n+1})\right]+\left[n_F(-M_{n+1})-n_F(-M_n)\right]}{(M_{n+1}-M_n)^2-(\omega+2i\Gamma)^2}\right. \\ &\times \frac{1}{M_{n+1}-M_n}+\left(1+\frac{\Delta^2}{M_n M_{n+1}}\right) \\ &\times \left.\frac{\left[n_F(-M_n)-n_F(M_{n+1})\right]+\left[n_F(-M_{n+1})-n_F(M_n)\right]}{(M_{n+1}+M_n)^2-(\omega+2i\Gamma)^2}\frac{1}{M_{n+1}+M_n}\right\},\end{aligned} \tag{9.3}$$

and

$$\begin{aligned}\sigma_0 &= -\frac{e^2 v_F^2 eB_0}{\pi}\sum_{n=0}^{\infty}\left(\left[n_F(M_n)-n_F(M_{n+1})\right]+\left[n_F(-M_{n+1})-n_F(-M_n)\right]\right) \\ &\times \left\{\left(1-\frac{\Delta^2}{M_n M_{n+1}}\right)\frac{1}{(M_{n+1}-M_n)^2-(\omega+2i\Gamma)^2}\right. \\ &\left.+\left(1+\frac{\Delta^2}{M_n M_{n+1}}\right)\frac{1}{(M_{n+1}+M_n)^2-(\omega+2i\Gamma)^2}\right\},\end{aligned} \tag{9.4}$$

where $n_F(\varepsilon)=1/\left\{1+\exp\left[(\varepsilon-\mu_c)/(k_B T)\right]\right\}$ is the Fermi–Dirac distribution, $v_F = 10^6$ m/s is the Fermi velocity, $M_n=\sqrt{\Delta^2+2n\hbar v_F^2|e\mathbf{B}_0|}$ is the energy of the nth Landau level. Δ is the excitonic bandgap that can be assumed zero for low magnetic fields, leading to the following simplified equations:

$$\sigma_{\rm d} = -\frac{ie^2(\omega + i2\Gamma)}{\pi\hbar^2}\left[\frac{1}{(\omega+2i\Gamma)^2}\int_0^\infty d\varepsilon\left(\frac{\partial n_F(\varepsilon)}{\partial\varepsilon} - \frac{\partial n_F(-\varepsilon)}{\partial\varepsilon}\right)\varepsilon - \int_0^\infty d\varepsilon \frac{n_F(-\varepsilon) - n_F(\varepsilon)}{(\omega+2i\Gamma)^2 - 4(\varepsilon/\hbar)^2}\right], \quad (9.5)$$

and

$$\sigma_0 = -\frac{e^2 v_F^2 eB_0}{\pi\hbar^2}\left[\frac{1}{(\omega+2i\Gamma)^2}\int_0^\infty d\varepsilon\left(\frac{\partial n_F(\varepsilon)}{\partial\varepsilon} - \frac{\partial n_F(-\varepsilon)}{\partial\varepsilon}\right) + \int_0^\infty d\varepsilon \frac{1}{(\omega+2i\Gamma)^2 - 4(\varepsilon/\hbar)^2}\right]. \quad (9.6)$$

The off-diagonal term of conductivity in Eq. (9.5) vanishes by assuming no magnetic bias field $B_0 = 0$, while the diagonal term $\sigma_{\rm d}$ in Eq. (9.6) is independent of the DC magnetic field.

Graphene's optical conductivity is divided into the intraband and interband parts, which correspond to free carrier absorption and transition from the valance band to the conduction band, respectively. In Eq. (9.5), the first term is due to the intraband contributions and the second term is related to the interband transitions. Equation (9.7) is the Drude-like analytical expression of the intraband transitions term [51, 61], while Eq. (9.8) is the approximate analytical expression obtained for the interband term when $k_B T << |\mu_c|$ and $k_B T << \hbar\omega$ [62]:

$$\sigma_{\rm d,intra} = i\frac{e^2 k_B T}{\pi\hbar^2(\omega+2i\Gamma)}\left[\frac{\mu_c}{k_B T} + 2\ln(e^{-\frac{\mu_c}{k_B T}} + 1)\right] \quad (9.7)$$

$$\sigma_{\rm d,inter} \approx \frac{ie^2}{4\pi\hbar}\ln\left(\frac{2|\mu_c| - (\omega+2i\Gamma)\hbar}{2|\mu_c| + (\omega+2i\Gamma)\hbar}\right) \quad (9.8)$$

The scattering rate Γ has an inverse relation to the relaxation time τ as $2\Gamma = \tau^{-1}$. The relaxation time varies with the Fermi energy (chemical potential) for a fixed Fermi velocity of graphene as $\tau = \mu_c / e v_F^2$. It is worth mentioning that the high ambient temperature of thermal emitters invalidates the condition for the closed-form approximation in Eq. (9.8). In our study, the general form in Eq. (9.5) is, therefore, evaluated numerically to obtain the exact value of the refractive index of graphene. However, Eq. (9.8) shows the relation of parameters to the interband contribution. For instance, this equation implies that $\sigma_{\mathrm{d,inter}}$ is purely imaginary when $2|\mu_c| > \hbar\omega$ and $\Gamma \approx 0$, or complex when $2|\mu_c| < \hbar\omega$.

The contributions of intraband and interband transitions in the optical conductivity significantly depend on the carrier density, so that each part has different strength at different frequency ranges. These contributions are also directly related to the chemical potential in graphene. Figure 9.4 describes the optical conductivity of graphene and its equivalent refractive index as a function of the chemical potential and the wavelength of incident light. For undoped suspended graphene (i.e., there is no electric field to tune the chemical potential), the interband transitions are responsible for the ~2.3% broadband absorption. The DC electric field bias tunes the chemical potential in graphene, adding either electrons or holes to the system. In this scenario, the interband transitions of electrons occur only for $2E_0 = \hbar\omega > 2E_F$, while other transitions are forbidden or blocked for $E_0 < 2E_F$, as shown in Fig. 9.4a. The absorption due to the interband transition is reduced by Pauli blocking, because the vacant states in conduction band are all occupied when the pumping light is intense enough for a specific relaxation process [37]. The intraband transitions are mainly responsible for the absorption in the far-infrared and also contribute, to some extent, to the mid-infrared optical response [63]. For short wavelengths in the visible range, graphene's optical conductivity is dominated by interband transitions [63]. The contribution of intraband transition decreases by increasing the chemical potential in the visible range and has, therefore, no significant effect on graphene's optical conductivity at these wavelengths.

Finally, the dielectric permittivity of monolayer graphene ε_G at optical frequencies is given by [48, 64]

$$\varepsilon_G = \frac{i\sigma_d}{\omega\varepsilon_0 t_G}, \tag{9.9}$$

where σ_d is the conductivity of graphene, t_G is the thickness of a single graphene layer, ω is the angular frequency, and ε_0 is the free-space electric permittivity [65]. Figures 9.4b and 9.4c show the real and imaginary parts of graphene's refractive index as a function of the chemical potential in graphene. It can be observed that in the visible range, the real and imaginary parts of graphene's refractive index are nearly independent of the chemical potential.

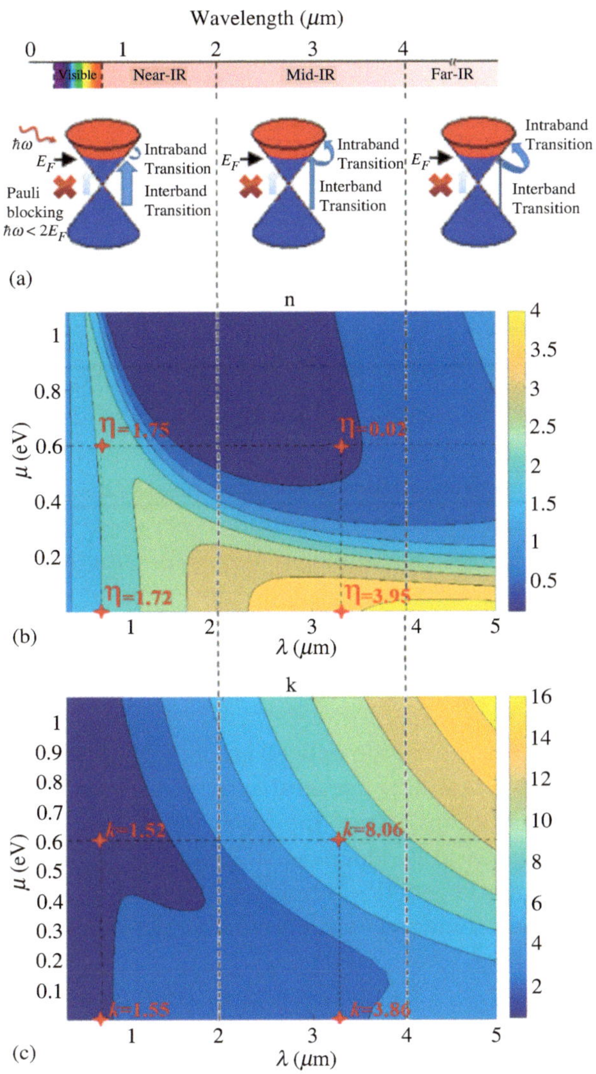

Figure 9.4 Sketch of graphene's dispersion for (a) doped graphene; interband transitions are forbidden for $\hbar\omega < 2E_F$ by the Pauli blocking. For visible light, the optical conductivity due to interband transitions is dominant. For near-infrared and mid-infrared radiations, the interband and intraband transitions have comparable contributions to the optical conductivity of graphene, while intraband transitions are dominant for far infrared. (b) The real part and (c) imaginary parts of the refractive index obtained by the Kubo formalism, as well as the equivalent changes in their values for visible and infrared radiation for chemical potential of μc = 0.0 eV and μc = 0.6 eV.

For longer wavelengths in the infrared range, the Drude-like part in Eq. (9.7) increases as the wavelength becomes longer, while the interband transition contribution significantly decreases, behaving as a step-like function with threshold $2|E_F|$ (small value when the photon energy is below $2|E_F|$, and large value when the photon energy is above $2|E_F|$). As such, the intraband transition contribution becomes comparable with the interband transition contribution, as shown in Fig. 9.4a. The control over intraband transitions can be obtained by tuning of the chemical potential in graphene, resulting in graphene applications in the infrared and THz ranges [31, 66, 67]. As such, graphene has been shown to be a good candidate for plasmonics and metamaterials in the mid-infrared [68, 69] and THz ranges [67, 70, 71]. For infrared radiation at a high ambient temperature of thermal emitter (e.g., 873 K in this chapter), the results show an order of magnitude decrease in the contribution of interband transitions, leading to larger contribution of intraband transitions to the total optical conductivity of graphene.

9.5 Summary

In this chapter, we have introduced graphene photonics properties, followed by presenting several graphene nanostructures for photonic applications. Then, the Kubo formalism is presented to calculate the optical conductivity of graphene. This model will be applied in a hybrid genetic optimization method to find a graphene-based aperiodic multilayer structure in Chapter 10.

References

1. Mak, K. F., Ju, L., Wang, F., and Heinz, T. F. (2012). Optical spectroscopy of graphene: From the far infrared to the ultraviolet, *Solid State Communications*, **152**, pp. 1341–1349.
2. Xia, F., Yan, H., and Avouris, P. (2013). The interaction of light and graphene: Basics, devices, and applications, *Proceedings of the IEEE*, **101**, pp. 1717–1731.
3. Eda, G. and Maier, S. A. (2013). Two-dimensional crystals: Managing light for optoelectronics, *ACS Nano*, **7**, pp. 5660–5665.

4. Xia, F., Wang, H., Xiao, D., Dubey, M., and Ramasubramaniam, A. (2014). Two-dimensional material nanophotonics, *Nature Photonics*, **8**, pp. 899–907.

5. Liu, Z., Song, L., Zhao, S., Huang, J., Ma, L., Zhang, J., Lou, J., and Ajayan, P. M. (2011). Direct growth of graphene/hexagonal boron nitride stacked layers, *Nano Letters*, **11**, pp. 2032–2037.

6. Mak, K. F., Lee, C., Hone, J., Shan, J., and Heinz, T. F. (2010). Atomically thin MoS_2: A new direct-gap semiconductor, *Physical Review Letters*, **105**, p. 136805.

7. Splendiani, A., Sun, L., Zhang, Y., Li, T., Kim, J., Chim, C.-Y., Galli, G., and Wang, F. (2010). Emerging photoluminescence in monolayer MoS_2, *Nano Letters*, **10**, pp. 1271–1275.

8. Xia, F., Wang, H., and Jia, Y. (2014). Rediscovering black phosphorus as an anisotropic layered material for optoelectronics and electronics, *Nature Communications*, **5**, p. 4458.

9. Buscema, M., Groenendijk, D. J., Blanter, S. I., Steele, G. A., van der Zant, H. S., and Castellanos-Gomez, A. (2014). Fast and broadband photoresponse of few-layer black phosphorus field-effect transistors, *Nano Letters*, **14**, pp. 3347–3352.

10. Low, T., Rodin, A., Carvalho, A., Jiang, Y., Wang, H., Xia, F., and Neto, A. C. (2014). Tunable optical properties of multilayer black phosphorus thin films, *Physical Review B*, **90**, p. 075434.

11. Nair, R. R., Blake, P., Grigorenko, A. N., Novoselov, K. S., Booth, T. J., Stauber, T., Peres, N. M., and Geim, A. K. (2008). Fine structure constant defines visual transparency of graphene, *Science*, **320**, pp. 1308–1308.

12. Kuzmenko, A., Van Heumen, E., Carbone, F., and Van Der Marel, D. (2008). Universal optical conductance of graphite, *Physical Review Letters*, **100**, p. 117401.

13. Kumar, A. and Zhou, C. (2010). The race to replace tin-doped indium oxide: Which material will win? *ACS Nano*, **4**, pp. 11–14.

14. Ellmer, K. (2012). Past achievements and future challenges in the development of optically transparent electrodes, *Nature Photonics*, **6**, pp. 809–817.

15. Han, T.-H., Lee, Y., Choi, M.-R., Woo, S.-H., Bae, S.-H., Hong, B. H., Ahn, J.-H., and Lee, T.-W. (2012). Extremely efficient flexible organic light-emitting diodes with modified graphene anode, *Nature Photonics*, **6**, pp. 105–110.

16. Bae, S., Kim, S. J., Shin, D., Ahn, J.-H., and Hong, B. H. (2012). Towards industrial applications of graphene electrodes, *Physica Scripta*, **2012**, p. 014024.

17. Yan, H., Xia, F., Zhu, W., Freitag, M., Dimitrakopoulos, C., Bol, A. A., Tulevski, G., and Avouris, P. (2011). Infrared spectroscopy of wafer-scale graphene, *ACS Nano*, **5**, pp. 9854–9860.

18. Bregar, V. B. (2004). Advantages of ferromagnetic nanoparticle composites in microwave absorbers, *IEEE Transactions on Magnetics*, **40**, pp. 1679–1684.

19. Han, Z., Li, D., Wang, H., Liu, X., Li, J., Geng, D., and Zhang, Z. (2009). Broadband electromagnetic-wave absorption by FeCo/C nanocapsules, *Applied Physics Letters*, **95**, p. 023114.

20. Novoselov, K. S., Fal, V., Colombo, L., Gellert, P., Schwab, M., and Kim, K. (2012). A roadmap for graphene, *Nature*, **490**, pp. 192–200.

21. Ishibashi, T., Furuta, T., Fushimi, H., Kodama, S., Nagatsuma, T., Shimizu, N., and Miyamoto, Y. (2000). InP/InGaAs uni-traveling-carrier photodiodes, *IEICE Transactions on Electronics*, **83**, pp. 938–949.

22. Xia, F., Mueller, T., Lin, Y.-M., Valdes-Garcia, A., and Avouris, P. (2009). Ultrafast graphene photodetector, *Nature Nanotechnology*, **4**, pp. 839–843.

23. Mueller, T., Xia, F., and Avouris, P. (2010). Graphene photodetectors for high-speed optical communications, *Nature Photonics*, **4**, pp. 297–301.

24. Echtermeyer, T., Britnell, L., Jasnos, P., Lombardo, A., Gorbachev, R., Grigorenko, A., Geim, A., Ferrari, A., and Novoselov, K. (2011). Strong plasmonic enhancement of photovoltage in graphene, *Nature Communications*, **2**, p. 458.

25. Kim, K., Choi, J.-Y., Kim, T., Cho, S.-H., and Chung, H.-J. (2011). A role for graphene in silicon-based semiconductor devices, *Nature*, **479**, pp. 338–344.

26. Engel, M., Steiner, M., Lombardo, A., Ferrari, A. C., Löhneysen, H. V., Avouris, P., and Krupke, R. (2012). Light–matter interaction in a microcavity-controlled graphene transistor, *Nature Communications*, **3**, p. 906.

27. Furchi, M., Urich, A., Pospischil, A., Lilley, G., Unterrainer, K., Detz, H., Klang, P., Andrews, A. M., Schrenk, W., and Strasser, G. (2012). Microcavity-integrated graphene photodetector, *Nano Letters*, **12**, pp. 2773–2777.

28. Reed, G. T., Mashanovich, G., Gardes, F., and Thomson, D. (2010). Silicon optical modulators, *Nature Photonics*, **4**, pp. 518–526.

29. Tang, Y., Chen, H.-W., Jain, S., Peters, J. D., Westergren, U., and Bowers, J. E. (2011). 50 Gb/s hybrid silicon traveling-wave electroabsorption modulator, *Optics Express*, **19**, pp. 5811–5816.

30. Wang, F., Zhang, Y., Tian, C., Girit, C., Zettl, A., Crommie, M., and Shen, Y. R. (2008). Gate-variable optical transitions in graphene, *Science*, **320**, pp. 206–209.

31. Liu, M., Yin, X., Ulin-Avila, E., Geng, B., Zentgraf, T., Ju, L., Wang, F., and Zhang, X. (2011). A graphene-based broadband optical modulator, *Nature*, **474**, pp. 64–67.

32. Gan, X., Shiue, R.-J., Gao, Y., Mak, K. F., Yao, X., Li, L., Szep, A., Walker Jr, D., Hone, J., and Heinz, T. F. (2013). High-contrast electrooptic modulation of a photonic crystal nanocavity by electrical gating of graphene, *Nano Letters*, **13**, pp. 691–696.

33. Sensale-Rodriguez, B., Fang, T., Yan, R., Kelly, M. M., Jena, D., Liu, L., and Xing, H. (2011). Unique prospects for graphene-based terahertz modulators, *Applied Physics Letters*, **99**, p. 113104.

34. Tan, W., Su, C., Knize, R., Xie, G., Li, L., and Tang, D. (2010). Mode locking of ceramic Nd: yttrium aluminum garnet with graphene as a saturable absorber, *Applied Physics Letters*, **96**, p. 031106.

35. Keller, U., Weingarten, K. J., Kartner, F. X., Kopf, D., Braun, B., Jung, I. D., Fluck, R., Honninger, C., Matuschek, N., and Der Au, J. A. (1996). Semiconductor saturable absorber mirrors (SESAM's) for femtosecond to nanosecond pulse generation in solid-state lasers, *IEEE Journal of Selected Topics in Quantum Electronics*, **2**, pp. 435–453.

36. Bao, Q., Zhang, H., Wang, Y., Ni, Z., Yan, Y., Shen, Z. X., Loh, K. P., and Tang, D. Y. (2009). Atomic layer graphene as a saturable absorber for ultrafast pulsed lasers, *Advanced Functional Materials*, **19**, pp. 3077–3083.

37. Sun, Z., Hasan, T., Torrisi, F., Popa, D., Privitera, G., Wang, F., Bonaccorso, F., Basko, D. M., and Ferrari, A. C. (2010). Graphene mode-locked ultrafast laser, *ACS Nano*, **4**(2), pp. 803–810.

38. Zhang, H., Tang, D., Knize, R., Zhao, L., Bao, Q., and Loh, K. P. (2010). Graphene mode locked, wavelength-tunable, dissipative soliton fiber laser, *Applied Physics Letters*, **96**, p. 111112.

39. Xu, J.-L., Li, X.-L., He, J.-L., Hao, X.-P., Wu, Y.-Z., Yang, Y., and Yang, K.-J. (2011). Performance of large-area few-layer graphene saturable absorber in femtosecond bulk laser, *Applied Physics Letters*, **99**, p. 261107.

40. De Souza, E., Nuss, M., Knox, W., and Miller, D. (1995). Wavelength-division multiplexing with femtosecond pulses, *Optics Letters*, **20**, pp. 1166–1168.

41. Drexler, W. (2004). Ultrahigh-resolution optical coherence tomography, *Journal of Biomedical Optics*, **9**, pp. 47–74.

42. Liu, X., Du, D., and Mourou, G. (1997). Laser ablation and micromachining with ultrashort laser pulses, *IEEE Journal of Quantum Electronics*, **33**, pp. 1706–1716.

43. Rana, F. (2008). Graphene terahertz plasmon oscillators, *IEEE Transactions on Nanotechnology*, **7**, pp. 91–99.

44. Bao, Q., Zhang, H., Wang, B., Ni, Z., Lim, C. H. Y. X., Wang, Y., Tang, D. Y., and Loh, K. P. (2011). Broadband graphene polarizer, *Nature Photonics*, **5**, pp. 411–415.

45. Stauber, T., Peres, N., and Geim, A. (2008). Optical conductivity of graphene in the visible region of the spectrum, *Physical Review B*, **78**, p. 085432.

46. Stauber, T., Noriega-Pérez, D., and Schliemann, J. (2015). Universal absorption of two-dimensional systems, *Physical Review B*, **91**, p. 115407.

47. Iorsh, I. V., Mukhin, I. S., Shadrivov, I. V., Belov, P. A., and Kivshar, Y. S. (2013). Hyperbolic metamaterials based on multilayer graphene structures, *Physical Review B*, **87**, p. 075416.

48. Falkovsky, L. and Pershoguba, S. (2007). Optical far-infrared properties of a graphene monolayer and multilayer, *Physical Review B*, **76**, p. 153410.

49. Gusynin, V. and Sharapov, S. (2006). Transport of Dirac quasiparticles in graphene: Hall and optical conductivities, *Physical Review B*, **73**, p. 245411.

50. Gusynin, V., Sharapov, S., and Carbotte, J. (2006). Unusual microwave response of Dirac quasiparticles in graphene, *Physical Review Letters*, **96**, p. 256802.

51. Gusynin, V., Sharapov, S., and Carbotte, J. (2006). Magneto-optical conductivity in graphene, *Journal of Physics: Condensed Matter*, **19**, p. 026222.

52. Vakil, A. and Engheta, N. (2011). Transformation optics using graphene, *Science*, **332**, pp. 1291–1294.

53. Ando, T., Zheng, Y., and Suzuura, H. (2002). Dynamical conductivity and zero-mode anomaly in honeycomb lattices, *Journal of the Physical Society of Japan*, **71**, pp. 1318–1324.

54. Wunsch, B., Stauber, T., Sols, F., and Guinea, F. (2006). Dynamical polarization of graphene at finite doping, *New Journal of Physics*, **8**, pp. 318.

55. Falkovsky, L. and Varlamov, A. (2007). Space–time dispersion of graphene conductivity, *The European Physical Journal B-Condensed Matter and Complex Systems*, **56**, pp. 281–284.

56. Hanson, G. W. (2008). Dyadic Green's functions for an anisotropic, non-local model of biased graphene, *IEEE Transactions on Antennas and Propagation*, **56**, pp. 747–757.

57. Wehling, T., Novoselov, K., Morozov, S., Vdovin, E., Katsnelson, M., Geim, A., and Lichtenstein, A. (2008). Molecular doping of graphene, *Nano Letters*, **8**, pp. 173–177.

58. Novoselov, K. S., Geim, A. K., Morozov, S. V., Jiang, D., Zhang, Y., Dubonos, S. V., Grigorieva, I. V., and Firsov, A. A. (2004). Electric field effect in atomically thin carbon films, *Science*, **306**, pp. 666–669.

59. Depine, R. A. (2016). Electromagnetics of graphene, in: *Graphene Optics: Electromagnetic Solution of Canonical Problems*, Morgan & Claypool Publishers.

60. Vakil, A. (2012). Transformation optics using graphene: One-atom-thick optical devices based on graphene. PhD Dissertation, University of Pennsylvania, https://repository.upenn.edu/edissertations/715.

61. Hanson, G. W. (2008). Dyadic Green's functions and guided surface waves for a surface conductivity model of graphene, *Journal of Applied Physics*, **103**, p. 064302.

62. Gusynin, V., Sharapov, S., and Carbotte, J. (2007). Sum rules for the optical and Hall conductivity in graphene, *Physical Review B*, **75**, p. 165407.

63. Dawlaty, J. M., Shivaraman, S., Strait, J., George, P., Chandrashekhar, M., Rana, F., Spencer, M. G., Veksler, D., and Chen, Y. (2008). Measurement of the optical absorption spectra of epitaxial graphene from terahertz to visible, *Applied Physics Letters*, **93**, p. 131905.

64. Wang, H., Yang, Y., and Wang, L. (2015). Infrared frequency-tunable coherent thermal sources, *Journal of Optics*, **17**, p. 045104.

65. Sharifi, S., Banadaki, Y. M., You, C., Lorenzo, S., Veronis, G., and Dowling, J. P. (2017). Aperiodic multilayer graphene based tunable and switchable thermal emitter at mid-infrared frequencies, *ECS Meeting Abstracts*, **1367**.

66. Sensale-Rodriguez, B., Yan, R., Kelly, M. M., Fang, T., Tahy, K., Hwang, W. S., Jena, D., Liu, L., and Xing, H. G. (2012). Broadband graphene terahertz

modulators enabled by intraband transitions, *Nature Communications*, **3**, p. 780.

67. Andryieuski, A. and Lavrinenko, A. V. (2013). Graphene metamaterials based tunable terahertz absorber: Effective surface conductivity approach, *Optics Express*, **21**, pp. 9144–9155.
68. Jablan, M., Buljan, H., and Soljačić, M. (2009). Plasmonics in graphene at infrared frequencies, *Physical Review B*, **80**, p. 245435.
69. Mousavi, S. H., Kholmanov, I., Alici, K. B., Purtseladze, D., Arju, N., Tatar, K., Fozdar, D. Y., Suk, J. W., Hao, Y., and Khanikaev, A. B. (2013). Inductive tuning of Fano-resonant metasurfaces using plasmonic response of graphene in the mid-infrared, *Nano Letters*, **13**, pp. 1111–1117.
70. Zhang, Y., Feng, Y., Zhu, B., Zhao, J., and Jiang, T. (2014). Graphene-based tunable metamaterial absorber and polarization modulation in terahertz frequency, *Optics Express*, **22**, pp. 22743–22752.
71. Lin, I.-T., Lai, Y.-P., Wu, K.-H., and Liu, J.-M. (2014). Terahertz optoelectronic property of graphene: Substrate-induced effects on plasmonic characteristics, *Applied Sciences*, **4**, pp. 28–41.

Chapter 10

Graphene-Based Thermal Emitter

10.1 Introduction

At finite temperatures, all materials emit electromagnetic radiation due to the thermally induced motion of particles and quasiparticles [1]. A perfect thermal emitter follows Planck's law of blackbody radiation, which is broadband, incoherent, and isotropic, with a spectral profile and intensity that are dependent on the emissivity of a material, and that vary only with changes in temperature. However, it is desirable to have an emitter that radiates only within a certain frequency bandwidth, i.e., a selective emitter, for various applications, including infrared sensing, thermal imaging, and thermophotovoltaics. The spectral features of the controlled thermal emission (e.g., wavelength, bandwidth, peak emissivity, and angular characteristics) are strongly dependent on the choice of both materials and structures of the emitters.

The thermal emittance of both bulk materials and multilayer structures has been widely investigated over recent years. The thermal radiation from bulk materials is characterized by incoherent, isotropic, and broadband radiation spectra that vary depending on the properties of the selected material and its temperature. The thermal radiation spectra can be drastically altered by utilizing textured surfaces [2] and multilayer structures [3]. Highly directional and quasi-coherent thermal radiation sources have many important

Graphene Nanostructures: Modeling, Simulation, and Applications in Electronics and Photonics
Yaser M. Banadaki and Safura Sharifi

ISBN 978-981-4800-36-5 (Hardcover), 978-0-429-02221-0 (eBook)
www.jennystanford.com

applications as antennae [4], lasers [5, 6], and in photovoltaics [7, 8]. It has been shown that the directionality and coherence of blackbody emission can be controlled by nanoengineered structures such as patterned gratings [9, 10], photonic crystals [11, 12], microcavity resonators [4, 13–15], metasurfaces [1, 16, 17], and graphene nanostructures [18, 19]. Using the appropriate design of optical nanostructures, arbitrary shaping of thermal emission spectra can be realized, from single-peak ultra-narrowband emission for mid-infrared sensing [20] to a stepwise emissivity spectrum for thermophotovoltaics [21]. Coherent infrared thermal emitters with tunable emitting frequencies in a broad spectral range are highly desired for numerous promising applications in energy harvesting [15, 22, 23], chemical sensing [24], infrared sources [25], thermal circuits [18, 26], and radiative cooling [27].

Graphene, an atomic layer of carbon, has excellent electronic, photonic, mechanical, and thermal properties. It provides a unique platform for controlling thermal emittance at infrared wavelengths. In this chapter, we examine how graphene's optical properties can be employed to design a high-performance thermal emitter. We demonstrate that the graphene-based aperiodic multilayer structure enables the design of new narrowband, tunable, and switchable thermal infrared emitters, tailoring the radiation spectra of bulk materials [7]. To simulate the structure and optimize the layer thicknesses and materials, we use the genetic global optimization algorithm coupled to a transfer matrix code as explained in Chapter 9. We investigate the effect of the chemical potential of graphene layers on the tunability and switchability of the aperiodic multilayer structure.

10.2 Thermal Emitter and Blackbody Radiation

The electromagnetic spectrum is the range of all possible wavelengths, ranging from gamma ray radiation down to radio wave radiation, as shown in Fig. 10.1. All objects that are warmer than their surroundings emit photons within a certain range of wavelengths. This range is not a characteristic of the materials in the object and exclusively depends on the surface temperature of the object that is emitting the electromagnetic radiation. As such, gamma radiation and X-rays are emitted by very hot objects of temperature of a million kelvin, while photons with longer wavelengths in the infrared and radio wavelength range are emitted by cooler objects. The higher the temperature, the more intense the emitted radiation. Also the wavelength of maximum emission shifts to shorter wavelengths by increasing the temperature of the object.

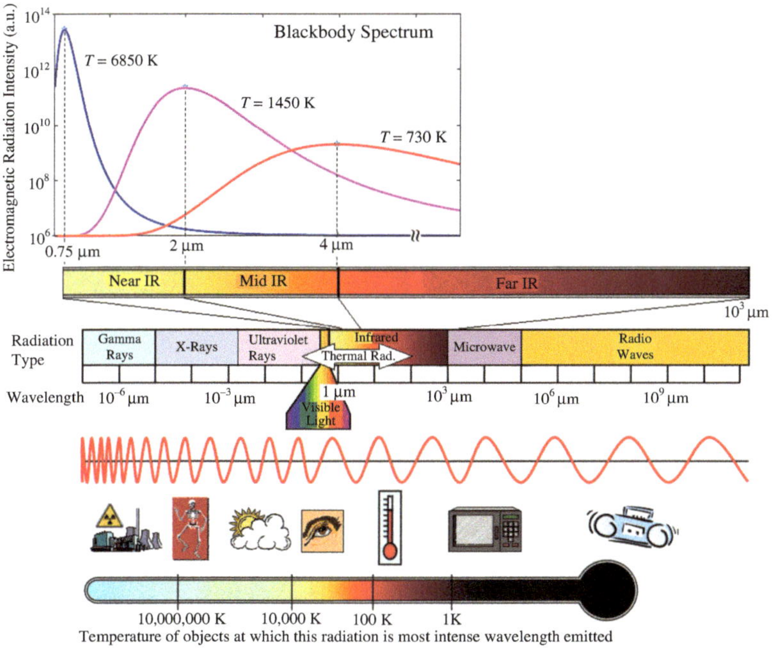

Figure 10.1 Electromagnetic spectrum as well as the thermal radiation spectrum, including the blackbody radiation at near-infrared and mid-infrared wavelengths.

To find the power radiated per unit area and wavelength, one can calculate the Planck blackbody spectrum (in units of watts per square meter per nanometer) as follows:

$$B(\lambda,T)=\frac{2hc^2}{\lambda^5}\frac{1}{e^{\frac{hc}{\lambda k_B T}}-1} \tag{10.1}$$

where k_B is the Boltzmann constant, h is Planck's constant, c is the speed of light, T is the ambient temperature, and λ is the wavelength.

An idealized blackbody absorbs all radiation that falls into the range enforced by the temperature of the object. Kirchhoff's law of thermal radiation states that the emittance and absorptance of an object must be equal for a given frequency, direction, and polarization under the condition of thermal equilibrium. As such, the normalized power radiated per unit area and unit wavelength by a non-blackbody in the normal direction as a function of wavelength and temperature can be calculated as follows:

$$\bar{\mu}(\lambda)=\frac{\in_{\text{Total}}(\lambda)\,B(\lambda,T)}{\max\limits_{\lambda}[B(\lambda,T)]} \tag{10.2}$$

where $\in_{\text{Total}}(\lambda) = [\in_{\text{TE}}(\lambda) + \in_{\text{TM}}(\lambda)]/2$ is the averaged emittance from both transverse electric (TE) and transverse magnetic (TM) polarizations in the normal direction. The value of $\bar{\mu}(\lambda)$ indicates how well the multilayer structure emits photons at a given wavelength in the normal direction.

10.3 Tunable Narrowband Thermal Emitters

The basic method to achieve narrowband absorptivity (emissivity) is the direct use of rare earth oxides, which inherently cause strong absorption at fixed wavelengths [28, 29]. However, the arbitrary control of an emission wavelength or an emission bandwidth is limited to the properties of chosen materials that determine the magnitude of interaction between light and materials. A selective emitter can be achieved by photonic bandgaps in photonic crystals composed of metallic and dielectric structures [9, 30]. As electromagnetic fields are strongly attenuated below the plasma frequency of metals [31, 32], they introduce more flexibility in creating a thermal emitter with broadband frequency selectivity [6, 33]. In addition, metals are potentially suitable for near-infrared selective thermal emitters, since they have large absorption in the near-infrared frequencies with stable properties at high temperatures. However, metals also have high reflectivity in mid- and far-infrared frequencies and consequently structures composed of metals can potentially show low emissivity [34]. As such, the surface is required to be modified periodically by an array of grooves [6] or holes [33] to enhance emission at infrared frequencies. In this way, the radiative and absorptive rates of the photonic crystal resonances dictate the emissivity spectrum so that the geometrical parameters enable a broadband, selective thermal emitter.

A narrowband thermal emission can be achieved using metallic nanostructures so that the optical resonant modes, confined in the so-called Salisbury screen [35] and Fabry–Pérot cavity [36, 37], are excited on the metal surface, leading to enhanced absorptivity (emissivity) at those resonant wavelengths [12, 17, 21, 38–41]. According to the Purcell effect [42], thermal radiation from an optical resonator can be dramatically modulated by the resonance mode designed in the infrared range, leading to narrowband thermal emission at the resonant frequency. Liu et al. [43] demonstrated that the matched mode of the emitter can be lost when the resonance mode is electrically quasi-static, i.e., the electric field oscillates in phase, resulting in the fundamental limit of the spectral thermal emission power from an optical resonator.

Metamaterials-based structures have also led to narrowband thermal emission [1, 44–50]. The effective permittivity and permeability of the entire structure are artificially controlled by combining sub-wavelength metallic elements with thin dielectric layers in a properly designed structure [51], leading to perfect absorption [9, 42, 52, 53] (maximum emission) at the resonant wavelengths. It should be noted, however, that strong free-carrier absorption due to metals leads to undesired emission over an extensive wavelength range together with the broadening of the emission peaks in selective thermal emitters designed by photonic crystals and metamaterials [54]. In addition, a narrowband resonance achieved in these structures cannot be changed dynamically to other operation frequencies due to the limitation in the properties and functionalities of the available conventional metals.

Infrared frequency-tunable coherent thermal emitters play an important role in material analysis providing noninvasive information on chemical composition and bonds. The dynamic control of thermal radiation has been demonstrated through in situ modification of material emissivity. This has been achieved with nanophotonic structures that incorporate phase change materials so that the emissivity can be electronically manipulated by controlling the charge injection, and consequently the polariton modes in the structure. Cong et al. [55] demonstrated that a tunable selective absorber can be designed by InSb, whose carrier density can be adjusted by utilizing optical pump or changing the surrounding temperature, altering the resonance frequency of split rings. Similarly, tunable perfect thermal emitters could be designed by the genesis of new materials as well as novel design of photonic and electronic structures.

Graphene is a zero-bandgap semimetal with high carrier mobility at room temperature, which allows strong interaction with terahertz (THz) and mid-infrared waves [56, 57]. The propagation of these waves can be actively controlled by varying the chemical potential[1] in graphene, which can be tuned by chemical doping, voltage bias, external magnetic field, or optical excitation [58]. However, graphene has low single-pass optical absorption due to its single-atom-layer thickness and thus the total absorption can be only achieved by novel designs of graphene-based nanostructures. Thongrattanasiri et al. [19] demonstrated tunable perfect absorbers with graphene ribbon array on dielectric spacer and metallic substrate [19, 59]. Wang et al. [60] demonstrated an infrared frequency-tunable selective thermal emitter made of graphene-covered silicon carbide (SiC) grating whose resonance frequency can be dynamically tuned by ~8.5% by varying graphene's chemical potential. Fang et al. [61] demonstrated tunable selective absorption in graphene disk arrays.

10.4 Graphene-Based Aperiodic Multilayer Structure

Figure 10.2 shows the schematic of the aperiodic graphene-based multilayer structure composed of alternating layers of graphene and hexagonal boron nitride (h-BN) insulator. In many cases, a multilayer structure may provide spectra-altering properties similar to that of more complex and harder-to-fabricate two-dimensional (2D) or 3D structures [7]. h-BN is an insulating layer with atomic thickness similar to graphene; thereby accurate control of the spacing between the graphene layers can be achieved in the aperiodic structures [62].

---

[1]The chemical potential and Fermi level correspond to the same physical quantity in material science so that the Fermi level E_F is the total chemical potential μ_c for electrons in the system. As both terms have been frequently used in the literature, these two terms are used interchangeably throughout this chapter.

The structure is sandwiched between two thick SiC layers, and a semi-infinite tungsten (W) layer is used as the substrate. Since the tungsten substrate is taken to be semi-infinite, the transmittance is identically zero, so that $A_{TE/TM}(\lambda) = 1 - R_{TE/TM}(\lambda)$, where $A_{TE/TM}(\lambda)$ is the TE/TM absorptance, $R_{TE/TM}(\lambda)$ is the TE/TM reflectance, and λ is the wavelength. The calculated absorptance can be equated to emittance because of Kirchhoff's second law and conservation of energy under thermal equilibrium.

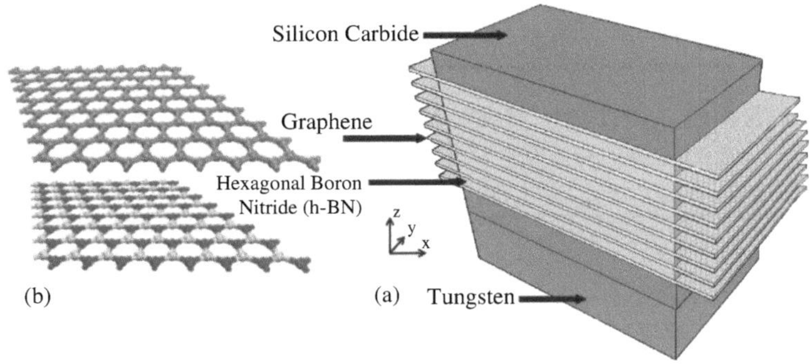

Figure 10.2 (a) Structure of the proposed thermal emitter. (b) The lattice structure of graphene and h-BN buffer layer.

The thicknesses of SiC layers and the number of h-BN layers can be altered and found by applying the optimization algorithm like genetic algorithm to achieve a structure with desired properties. The density of charge carriers associated with the chemical potential in graphene layers can be controlled by applying a DC bias field perpendicular to the graphene/h-BN surfaces [63]. The Kubo formula (Section 9.4) [64] is used to extract the optical conductivity and consequently the refractive index of graphene. For other materials such as h-BN, SiC, and W, the wavelength-dependent indices of refraction (both real and imaginary parts) are obtained from experimental data [65–67]. The effect of layer thickness variations due to thermal expansion on emittance/absorptance can be neglected. In addition, all materials used in the structure can tolerate high temperatures due to their high melting points [68].

Photonic nanostructures may be computationally designed to affect the motion of photons in a desired way so that a specific application is achieved. We find the optimum thicknesses of the layers in the aperiodic multilayer structure using a hybrid optimization method [69] consisting of a microgenetic global optimization algorithm coupled to a local optimization algorithm. Using this algorithm, we calculate the optimized thicknesses for maximizing the absorption to the perfect value of unity at a pre-specified wavelength and zero bias condition (μ = 0 eV) [70, 71]. The genetic algorithm is an iterative optimization procedure that starts with a randomly selected population of potential solutions [6]. We solve the electromagnetic equations (Maxwell's equations) in the structure using the transfer matrix method [72]. The field distribution in aperiodic multilayer structures is obtained, leading to information about the absorption, transmission, and reflection properties of the structure. While 2D and 3D structures are computationally intensive, 1D structures can be simulated to provide a proof of concept to design and implement more complex structures.

10.5 Selectivity, Tunability, and Switchability

The aperiodic multilayer structure is optimized to produce narrowband infrared thermal emission for both TE and TM polarizations. Thermal emission is considered at the ambient temperature of 873 K corresponding to the maximum emission of blackbody at the infrared range with the peak at λ = 3340 nm. This is the wavelength at which the genetic algorithm is applied to find the optimized structures with maximum absorptance. To seek the optimum structure, the chemical potentials of graphene layers are set equal to zero (μ_c = 0.0 eV), i.e., the chemical potentials are located at the Dirac point of graphene dispersion.

Figure 10.3a shows the aperiodic structure including the graphene layers (black lines) and layers of h-BN insulator. Utilizing the genetic optimization algorithm, each layer's thickness is selected independently so that the structure is completely aperiodic. Finally, utilizing the transfer matrix method [11], the absorptance, which is equal to the emittance, of the graphene-based structure is calculated. Figure 10.3b shows the normalized thermal emittance, $\overline{\mu}(\lambda)$, as a function of wavelength. We observe that the optimized structure exhibits almost perfect emittance of ~100% at λ = 3.34 μm. The narrowband perfect emittance is accomplished through interaction with the layered nanophotonic structure so that the bandwidth of 2.7 μm at T = 873 K reduces to 0.63 μm, showing more than four times narrower bandwidth compared to the blackbody radiation curve. It can be observed in Fig. 10.1 that the blackbody thermal emission is broadband, especially for lower ambient temperatures. The graphene-based heterostructure is an ultrathin nanophotonic structure that enables narrowband infrared emittance.

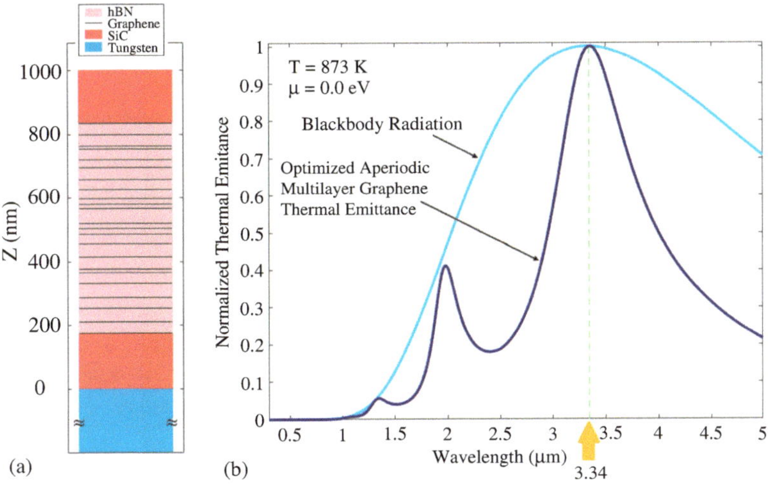

Figure 10.3 (a) Optimized structure with 23 layers of graphene and (b) normalized power radiated per unit area and unit wavelength by the structure in the normal direction.

The optimized thickness of the smallest h-BN layer is found to be d = 8.9 nm corresponding to 27 h-BN layers. The effect of this layer thickness on the emittance spectra of the optimized structure is investigated in Fig. 10.4. The thickness is changed to 5×d, 10×d, and 20×d, while the thicknesses of other layers are constant. This study can indicate the sensitivity of the optimized structure to fabrication process variations. Figure 10.3b shows the contribution of graphene layers and tungsten to the total power emitted from the proposed structure for different thicknesses. The SiC layers are not shown in the figure because they have less than 2% contribution to the total power regardless of the thickness of the smallest h-BN layer. At the optimized wavelength λ = 3.34 μm (dotted line), the tungsten and graphene layers contribute ~46% and ~52%, respectively, to the total power emitted from the optimized structure. The increase in the thickness of the thinnest h-BN layer leads to an increase in the reflectance, so that for the modified h-BN thickness of 20×*d*, the contributions of the tungsten and graphene layers are dramatically decreased to ~20% and ~25%, respectively. It can be observed that by increasing the thickness of the narrowest h-BN layer from the value that is obtained from the optimization process, the peak of radiated power shifts to longer wavelengths and decreases.

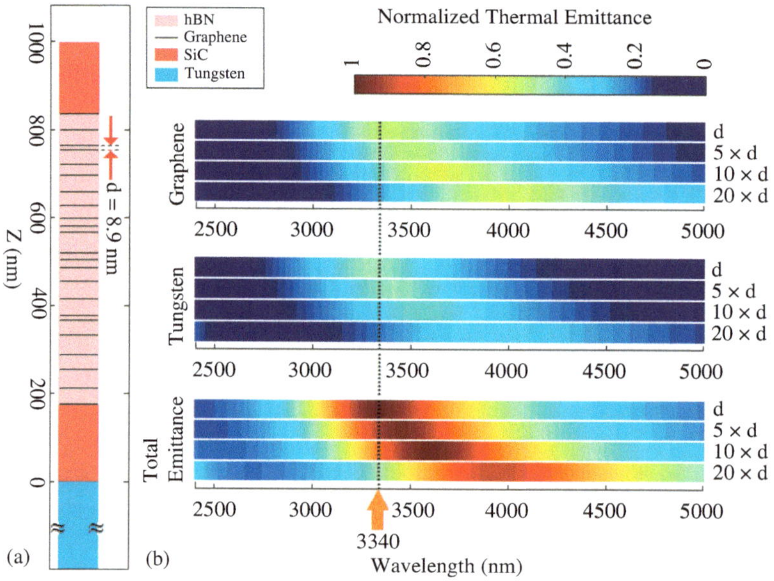

Figure 10.4 (a) Optimized structure with 23 layers of graphene including the location of the thinnest h-BN layer. (b) Effect of changing the optimized thickness of the thinnest h-BN layer on the emittance spectra, as well as on the contribution of graphene layers and tungsten to the total emittance.

The electric field in the structure can also be calculated for both TE and TM polarizations using the transfer matrix method [11]. Figure 10.5 shows the profile of the electric field amplitude normalized with respect to the field amplitude of the incident plane wave for the optimized structure in Fig. 10.3a. The structure is excited by a normally incident plane wave at $\lambda = 3.34$ μm, the wavelength at which the structures are optimized. The effect of varying the chemical potential of graphene can be clearly observed in the electric field profiles. For μ_c = 0.0 eV, at which the structure is optimized to achieve maximum absorptance, the electric field amplitude is almost flat in air. This suggests that the reflectance of the structure is almost zero, and the absorptance is, therefore, almost unity.

Figure 10.6 shows the partial absorptance and the corresponding reflectance as a function of wavelength for three different chemical potentials in graphene, μ_c = 0.0 eV, 0.4 eV, and 1.0 eV. At the chemical potential of μ_c = 0.0 eV, it is obvious that the reflectance is zero at the wavelength of 3.34 μm at which the structure is optimized, and perfect absorptance can be achieved. By increasing the chemical potential of graphene layers, the reflectance is increased and the corresponding absorptance on graphene layers is decreased at the wavelength at which the structure is optimized. As such at μ_c = 1.0 eV, tungsten is almost the only layer that contributes to the total absorption at the wavelength of 3.34 μm. At this wavelength, the absorptances are 1.0, 0.97, 0.82, 0.64, 0.48, and 0.35 for chemical potentials of μ_c = 0.0 eV, 0.2 eV, 0.4 eV, 0.6 eV, 0.8 eV, and 1.0 eV, respectively. This shows the possibility of switching, i.e., turning off, the absorptance of the proposed structure.

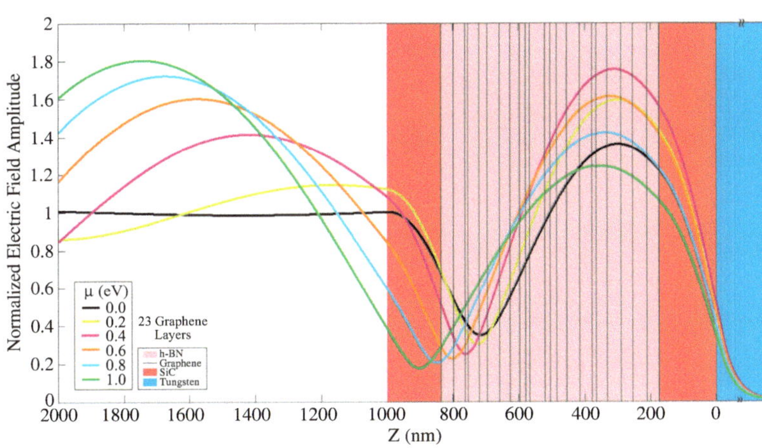

Figure 10.5 Optimized electric field amplitude normalized with respect to the field amplitude of the incident plane wave for the optimized structure with 23 graphene layers.

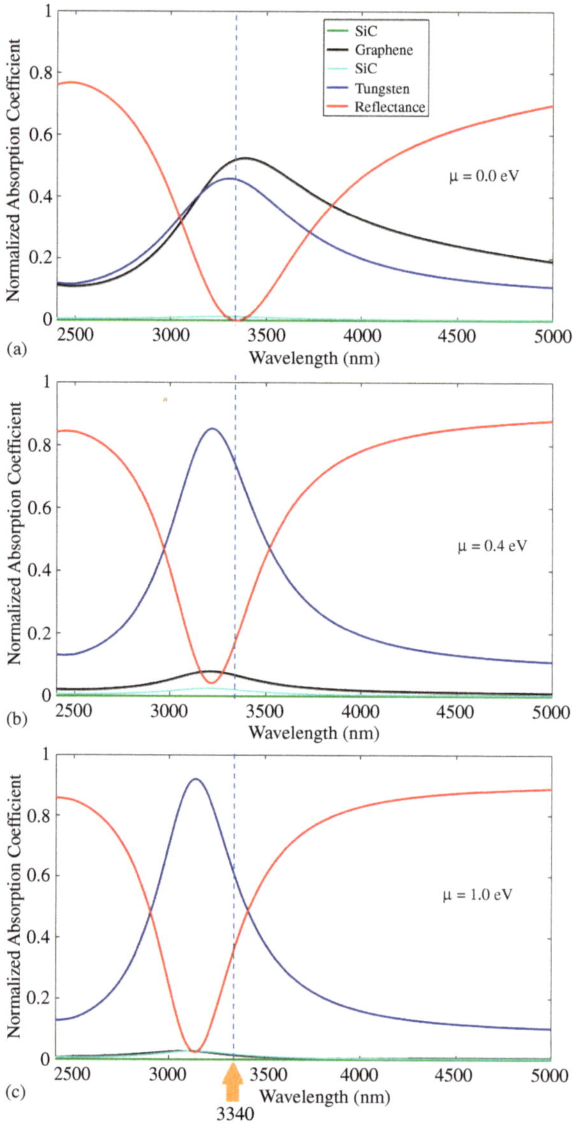

Figure 10.6 Effect of changing the chemical potential in graphene, i.e., (a) μ_c = 0.0 eV, (b) 0.4 eV, and (c) 1.0 eV, on the partial absorptance and the reflectance.

Figure 10.7 shows the effect of the chemical potential on the absorptance of the optimized structure. It can be observed that the maximum of the absorptance is close to unity but shifts to shorter wavelengths, as the chemical potential is varied. This shift is associated with the change in the chemical potential of graphene. In other words, the increase in the chemical potential reduces the absorption in graphene layers, but other layers and, especially, tungsten increase their contribution to the total absorption, so that the maximum absorption shifts to shorter wavelengths. This indicates that the proposed graphene-based aperiodic structure can be electrically controlled, enabling a tunable thermal emitter.

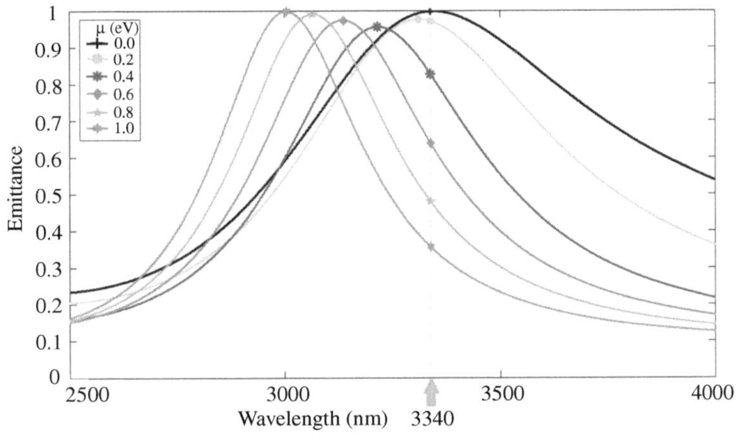

Figure 10.7 Absorptance of the optimized structure for different chemical potentials, μ_c = 0.0 eV, 0.2 eV, 0.4 eV, 0.6 eV, 0.8 eV, and 1.0 eV.

10.6 Summary

In this chapter, we have examined graphene optical properties in the design of a narrowband, tunable, and switchable thermal emitter. We have simulated the structure using a hybrid optimization method, consisting of a microgenetic global optimization algorithm coupled to a local optimization algorithm. We have investigated the effect of chemical potential of graphene layers on the tunability and switchability of the aperiodic multilayer structure. The optimized graphene-based structure exhibits narrowband perfect emittance of ~100% at λ = 3.34 μm. The bandwidth at T = 873 K reduces to 0.63 μm, resulting in more than four times narrower bandwidth compared to the blackbody radiation curve. We have also studied the sensitivity of the optimized structure to fabrication process variations. The absorptance on graphene layers is decreased, and the maximum absorption shifts to shorter wavelengths by increasing the chemical potential of graphene layers. This indicates that the proposed graphene-based aperiodic structure can be electrically controlled, enabling a tunable thermal emitter.

References

1. Liu, X., Tyler, T., Starr, T., Starr, A. F., Jokerst, N. M., and Padilla, W. J. (2011). Taming the blackbody with infrared metamaterials as selective thermal emitters, *Physical Review Letters*, **107**, p. 045901.
2. Ghebrebrhan, M., Bermel, P., Yeng, Y., Celanovic, I., Soljačić, M., and Joannopoulos, J. (2011). Tailoring thermal emission via Q matching of photonic crystal resonances, *Physical Review A*, **83**, p. 033810.
3. Rephaeli, E. and Fan, S. (2009). Absorber and emitter for solar thermo-photovoltaic systems to achieve efficiency exceeding the Shockley–Queisser limit, *Optics Express*, **17**, pp. 15145–15159.
4. Schuller, J. A., Taubner, T., and Brongersma, M. L. (2009). Optical antenna thermal emitters, *Nature Photonics*, **3**, pp. 658–661.
5. Marquier, F., Greffet, J.-J., Collin, S., Pardo, F., and Pelouard, J. (2005). Resonant transmission through a metallic film due to coupled modes, *Optics Express*, **13**, pp. 70–76.
6. Chan, D. L., Soljačić, M., and Joannopoulos, J. (2006). Thermal emission and design in 2D-periodic metallic photonic crystal slabs, *Optics Express*, **14**, pp. 8785–8796.
7. Fleming, J., Lin, S., El-Kady, I., Biswas, R., and Ho, K. (2002). All-metallic three-dimensional photonic crystals with a large infrared bandgap, *Nature*, **417**, pp. 52–55.
8. Florescu, M., Lee, H., Puscasu, I., Pralle, M., Florescu, L., Ting, D. Z., and Dowling, J. P. (2007). Improving solar cell efficiency using photonic band-gap materials, *Solar Energy Materials and Solar Cells*, **91**, pp. 1599–1610.
9. Greffet, J.-J., Carminati, R., Joulain, K., Mulet, J.-P., Mainguy, S., and Chen, Y. (2002). Coherent emission of light by thermal sources, *Nature*, **416**, pp. 61–64.
10. Han, S. and Norris, D. (2010). Beaming thermal emission from hot metallic bull's eyes, *Optics Express*, **18**, pp. 4829–4837.

11. Cornelius, C. M. and Dowling, J. P. (1999). Modification of Planck blackbody radiation by photonic band-gap structures, *Physical Review A*, **59**, p. 4736.

12. Lin, S.-Y., Moreno, J., and Fleming, J. (2003). Three-dimensional photonic-crystal emitter for thermal photovoltaic power generation, *Applied Physics Letters*, **83**, pp. 380–382.

13. Maruyama, S., Kashiwa, T., Yugami, H., and Esashi, M. (2001). Thermal radiation from two-dimensionally confined modes in microcavities, *Applied Physics Letters*, **79**, pp. 1393–1395.

14. Kusunoki, F., Takahara, J., and Kobayashi, I. (2003). Qualitative change of resonant peaks in thermal emission from periodic array of microcavities, *Electronics Letters*, **39**, pp. 23–24.

15. Sai, H., Kanamori, Y., and Yugami, H. (2003). High-temperature resistive surface grating for spectral control of thermal radiation, *Applied Physics Letters*, **82**, pp. 1685–1687.

16. Ikeda, K., Miyazaki, H., Kasaya, T., Yamamoto, K., Inoue, Y., Fujimura, K., Kanakugi, T., Okada, M., Hatade, K., and Kitagawa, S. (2008). Controlled thermal emission of polarized infrared waves from arrayed plasmon nanocavities, *Applied Physics Letters*, **92**, p. 021117.

17. Biener, G., Dahan, N., Niv, A., Kleiner, V., and Hasman, E. (2008). Highly coherent thermal emission obtained by plasmonic bandgap structures, *Applied Physics Letters*, **92**, p. 081913.

18. Liu, B., Liu, Y., and Shen, S. (2014). Thermal plasmonic interconnects in graphene, *Physical Review B*, **90**, p. 195411.

19. Thongrattanasiri, S., Koppens, F. H., and De Abajo, F. J. G. (2012). Complete optical absorption in periodically patterned graphene, *Physical Review Letters*, **108**, p. 047401.

20. Liu, N., Mesch, M., Weiss, T., Hentschel, M., and Giessen, H. (2010). Infrared perfect absorber and its application as plasmonic sensor, *Nano Letters*, **10**, pp. 2342–2348.

21. Chan, W. R., Bermel, P., Pilawa-Podgurski, R. C., Marton, C. H., Jensen, K. F., Senkevich, J. J., Joannopoulos, J. D., Soljačić, M., and Celanovic, I. (2013). Toward high-energy-density, high-efficiency, and moderate-temperature chip-scale thermophotovoltaics, *Proceedings of the National Academy of Sciences*, **110**, pp. 5309–5314.

22. Basu, S., Chen, Y. B., and Zhang, Z. (2007). Microscale radiation in thermophotovoltaic devices: A review, *International Journal of Energy Research*, **31**, pp. 689–716.

23. Wang, H. and Wang, L. (2013). Perfect selective metamaterial solar absorbers, *Optics Express*, **21**, pp. A1078–A1093.

24. Moelders, N., Pralle, M. U., McNeal, M. P., Puscasu, I., Last, L., Ho, W., Greenwald, A. C., Daly, J. T., and Johnson, E. A. (2002). Designing thermally uniform MEMs hot micro-bolometers, *MRS Proceedings*, **729**, U5.2. doi:10.1557/PROC-729-U5.2.

25. Brucoli, G., Bouchon, P., Haïdar, R., Besbes, M., Benisty, H., and Greffet, J.-J. (2014). High efficiency quasi-monochromatic infrared emitter, *Applied Physics Letters*, **104**, p. 081101.

26. Otey, C. R., Lau, W. T., and Fan, S. (2010). Thermal rectification through vacuum, *Physical Review Letters*, **104**, p. 154301.

27. Rephaeli, E., Raman, A., and Fan, S. (2013). Ultrabroadband photonic structures to achieve high-performance daytime radiative cooling, *Nano Letters*, **13**, pp. 1457–1461.

28. Sai, H., Yugami, H., Nakamura, K., Nakagawa, N., Ohtsubo, H., and Maruyama, S. (2000). Selective emission of $Al_2O_3/Er_3Al_5O_{12}$ eutectic composite for thermophotovoltaic generation of electricity, *Japanese Journal of Applied Physics*, **39**, p. 1957.

29. Bitnar, B., Durisch, W., Mayor, J.-C., Sigg, H., and Tschudi, H. (2002). Characterisation of rare earth selective emitters for thermophotovoltaic applications, *Solar Energy Materials and Solar Cells*, **73**, pp. 221–234.

30. Laroche, M., Carminati, R., and Greffet, J.-J. (2006). Coherent thermal antenna using a photonic crystal slab, *Physical Review Letters*, **96**, p. 123903.

31. Smith, D., Schultz, S., Kroll, N., Sigalas, M., Ho, K., and Soukoulis, C. (1994). Experimental and theoretical results for a two-dimensional metal photonic bandgap cavity, *Applied Physics Letters*, **65**, pp. 645–647.

32. Brown, E. and McMahon, O. (1995). Large electromagnetic stop bands in metallodielectric photonic crystals, *Applied Physics Letters*, **67**, pp. 2138–2140.

33. Celanovic, I., Jovanovic, N., and Kassakian, J. (2008). Two-dimensional tungsten photonic crystals as selective thermal emitters, *Applied Physics Letters*, **92**, p. 193101.

34. Reif, F. (2009). *Fundamentals of Statistical and Thermal Physics*, Waveland Press.

35. Laroche, M., Marquier, F., Carminati, R., and Greffet, J.-J. (2005). Tailoring silicon radiative properties, *Optics Communications*, **250**, pp. 316–320.

36. Wang, L., Lee, B., Wang, X., and Zhang, Z. (2009). Spatial and temporal coherence of thermal radiation in asymmetric Fabry–Pérot resonance cavities, *International Journal of Heat and Mass Transfer*, **52**, pp. 3024–3031.

37. Wang, L., Basu, S., and Zhang, Z. (2012). Direct measurement of thermal emission from a Fabry–Pérot cavity resonator, *Journal of Heat Transfer*, **134**, p. 072701.

38. Sai, H. and Yugami, H. (2004). Thermophotovoltaic generation with selective radiators based on tungsten surface gratings, *Applied Physics Letters*, **85**, pp. 3399–3401.

39. Pralle, M., Moelders, N., McNeal, M., Puscasu, I., Greenwald, A., Daly, J., Johnson, E., George, T., Choi, D., and El-Kady, I. (2002). Photonic crystal enhanced narrow-band infrared emitters, *Applied Physics Letters*, **81**, pp. 4685–4687.

40. Rinnerbauer, V., Yeng, Y. X., Chan, W. R., Senkevich, J. J., Joannopoulos, J. D., Soljačić, M., and Celanovic, I. (2013). High-temperature stability and selective thermal emission of polycrystalline tantalum photonic crystals, *Optics Express*, **21**, pp. 11482–11491.

41. Miyazaki, H., Ikeda, K., Kasaya, T., Yamamoto, K., Inoue, Y., Fujimura, K., Kanakugi, T., Okada, M., Hatade, K., and Kitagawa, S. (2008). Thermal emission of two-color polarized infrared waves from integrated plasmon cavities, *Applied Physics Letters*, **92**, p. 141114.

42. Liu, B., Gong, W., Yu, B., Li, P., and Shen, S. (2017). Perfect thermal emission by nanoscale transmission line resonators, *Nano Letters*, **17**, pp. 666–672.

43. Liu, B., Li, J., and Shen, S. (2017). Resonant thermal infrared emitters in near-and far-fields, *ACS Photonics*, **4**, pp. 1552–1557.

44. Puscasu, I. and Schaich, W. L. (2008). Narrow-band, tunable infrared emission from arrays of microstrip patches, *Applied Physics Letters*, **92**, p. 233102.

45. Mason, J., Smith, S., and Wasserman, D. (2011). Strong absorption and selective thermal emission from a midinfrared metamaterial, *Applied Physics Letters*, **98**, p. 241105.

46. Abbas, M. N., Cheng, C.-W., Chang, Y.-C., Shih, M.-H., Chen, H.-H., and Lee, S.-C. (2011). Angle and polarization independent narrow-band thermal emitter made of metallic disk on SiO_2, *Applied Physics Letters*, **98**, p. 121116.

47. Wu, C., Neuner III, B., John, J., Milder, A., Zollars, B., Savoy, S., and Shvets, G. (2012). Metamaterial-based integrated plasmonic absorber/

emitter for solar thermo-photovoltaic systems, *Journal of Optics*, **14**, p. 024005.

48. Molesky, S., Dewalt, C. J., and Jacob, Z. (2013). High temperature epsilon-near-zero and epsilon-near-pole metamaterial emitters for thermophotovoltaics, *Optics Express*, **21**, pp. A96–A110.

49. Wang, C.-M. and Feng, D.-Y. (2014). Omnidirectional thermal emitter based on plasmonic nanoantenna arrays, *Optics Express*, **22**, pp. 1313–1318.

50. Miyazaki, H., Kasaya, T., Iwanaga, M., Choi, B., Sugimoto, Y., and Sakoda, K. (2014). Dual-band infrared metasurface thermal emitter for CO_2 sensing, *Applied Physics Letters*, **105**, p. 121107.

51. Pendry, J. B., Holden, A. J., Robbins, D., and Stewart, W. (1999). Magnetism from conductors and enhanced nonlinear phenomena, *IEEE Transactions on Microwave Theory and Techniques*, **47**, pp. 2075–2084.

52. Aydin, K., Ferry, V. E., Briggs, R. M., and Atwater, H. A. (2011). Broadband polarization-independent resonant light absorption using ultrathin plasmonic super absorbers, *Nature Communications*, **2**, p. 517.

53. Streyer, W., Law, S., Rosenberg, A., Roberts, C., Podolskiy, V., Hoffman, A., and Wasserman, D. (2014). Engineering absorption and blackbody radiation in the far-infrared with surface phonon polaritons on gallium phosphide, *Applied Physics Letters*, **104**, p. 131105.

54. Inoue, T., De Zoysa, M., Asano, T., and Noda, S. (2015). Realization of narrowband thermal emission with optical nanostructures, *Optica*, **2**, pp. 27–35.

55. Cong, J., Yun, B., and Cui, Y. (2013). The ratio of the kinetic inductance to the geometric inductance: A key parameter for the frequency tuning of the THz semiconductor split-ring resonator, *Optics Express*, **21**, pp. 20363–20375.

56. Falkovsky, L. and Varlamov, A. (2007). Space-time dispersion of graphene conductivity, *The European Physical Journal B-Condensed Matter and Complex Systems*, **56**, pp. 281–284.

57. Falkovsky, L. and Pershoguba, S. (2007). Optical far-infrared properties of a graphene monolayer and multilayer, *Physical Review B*, **76**, p. 153410.

58. Novoselov, K. S., Fal, V., Colombo, L., Gellert, P., Schwab, M., and Kim, K. (2012). A roadmap for graphene, *Nature*, **490**, pp. 192–200.

59. Alaee, R., Farhat, M., Rockstuhl, C., and Lederer, F. (2012). A perfect absorber made of a graphene micro-ribbon metamaterial, *Optics Express*, **20**, pp. 28017–28024.

60. Wang, H., Yang, Y., and Wang, L. (2015). Infrared frequency-tunable coherent thermal sources, *Journal of Optics*, **17**, p. 045104.

61. Fang, Z., Wang, Y., Schlather, A. E., Liu, Z., Ajayan, P. M., García de Abajo, F. J., Nordlander, P., Zhu, X., and Halas, N. J. (2013). Active tunable absorption enhancement with graphene nanodisk arrays, *Nano Letters*, **14**, pp. 299–304.

62. Yan, H., Li, X., Chandra, B., Tulevski, G., Wu, Y., Freitag, M., Zhu, W., Avouris, P., and Xia, F. (2012). Tunable infrared plasmonic devices using graphene/insulator stacks, *Nature Nanotechnology*, **7**, pp. 330–334.

63. Mattiucci, N., D'Aguanno, G., Alu, A., Argyropoulos, C., Foreman, J., and Bloemer, M. (2012). Taming the thermal emissivity of metals: A metamaterial approach, *Applied Physics Letters*, **100**, p. 201109.

64. Gusynin, V., Sharapov, S., and Carbotte, J. (2006). Magneto-optical conductivity in graphene, *Journal of Physics: Condensed Matter*, **19**, p. 026222.

65. Singh, S., Potopowicz, J., Van Uitert, L., and Wemple, S. (1971). Nonlinear optical properties of hexagonal silicon carbide, *Applied Physics Letters*, **19**, pp. 53–56.

66. Fernández-Perea, M., Larruquert, J. I., Aznárez, J. A., Méndez, J. A., Vidal-Dasilva, M., Gullikson, E., Aquila, A., Soufli, R., and Fierro, J. (2007). Optical constants of electron-beam evaporated boron films in the 6.8–900 eV photon energy range, *JOSA A*, **24**, pp. 3800–3807.

67. Aksyutov, L. (1977). Temperature dependence of the optical constants of tungsten and gold, *Journal of Applied Spectroscopy*, **26**, pp. 656–660.

68. Yeng, Y. X., Ghebrebrhan, M., Bermel, P., Chan, W. R., Joannopoulos, J. D., Soljačić, M., and Celanovic, I. (2012). Enabling high-temperature nanophotonics for energy applications, *Proceedings of the National Academy of Sciences*, **109**, pp. 2280–2285.

69. Rahmat-Samii, Y. and Michielssen, E. (1999). Electromagnetic optimization by genetic algorithms, *Microwave Journal*, **42**, pp. 232–232.

70. Granier, C. H., Afzal, F. O., Min, C., Dowling, J. P., and Veronis, G. (2014). Optimized aperiodic highly directional narrowband infrared emitters, *JOSA B*, **31**, pp. 1316–1321.

71. Granier, C. H., Afzal, F. O., Lorenzo, S. G., Reyes Jr, M., Dowling, J. P., and Veronis, G. (2014). Optimized aperiodic multilayer structures for use as narrow-angular absorbers, *Journal of Applied Physics*, **116**, p. 243101.

72. Taflove, A. and Hagness, S. C. (2005). *Computational Electrodynamics: The Finite-Difference Time-Domain Method*, Artech House.

Index